Antique Brass
Identification and Values

Mary Frank Gaston

COLLECTOR BOOKS
A Division of Schroeder Publishing Co., Inc.

The current values in this book should be used only as a guide. They are not intended to set prices, which vary from one section of the country to another. Auction prices as well as dealer prices vary greatly and are affected by condition as well as demand. Neither the Author nor the Publisher assumes responsibility for any losses that might be incurred as a result of consulting this guide.

To Jerry and Jeremy

Contents

Brass and Its Production 10

Collecting Brass 18

Useful Terms 27

Lighting Implements 31

Fireplace Equipment 47

Kitchen Items 63

Tools and Instruments 87

Hardware Fixtures 124

Decorative and Personal Objects 137

Object Index 186

Price Guide 188

Bibliography 191

Acknowledgements

A book is rarely the sole effort of one person, and this one is certainly no exception to that rule. Many people were involved in the production of this book, and I would like to express my appreciation to them.

I thank my publisher, Bill Schroeder, for offering me the opportunity to put together a book on such an interesting topic as brass. It has been a most enjoyable and informative undertaking to be able to pursue my own interest in this subject. I thank him also for publishing the book in color which really adds so much to a book of this type.

I thank my husband, Jerry, for his fine photography of all the items illustrated. Jerry had to adapt to a new set-up for this book, and brass, because of its shiny nature, is not an easy subject for photography. I am very pleased with the results. As always, I thank him for his support and encouragement which is always one hundred percent!

Without the help of the following individuals, however, this book on brass would never have become a reality. These dealers and collectors so generously allowed me to photograph their brass. Furthermore, they provided much helpful information gained through their own research and experience. Disruption of their shop, show booth, or home was graciously permitted and assistance provided when needed. My special thanks to each and every one of the following:

Anteaque Tyme, Dallas, Texas
D.J. Blackburn, Waco, Texas
Bill and Toncy Brown, Granny Had It Antiques, Dallas, Texas
Helen Buchanan, The Copper Lamp, Dallas, Texas
Bob Campbell, The Porch, Big D Bazaar, Dallas, Texas
Lamar and Mabel Chambers, Let's Decorate, Dallas, Texas
The Depot Antiques, New Braunfels, Texas
Maizelle Dunlap, Bessie Mai's Antiques, Ft. Worth, Texas

Eclectic Ideas, Dallas, Texas
Asa and Sue Ellis, Antiques by Ellis, Arlington, Texas
Linda Eulich and Isabelle Young, M.I.L.E. Galleries, Dallas, Texas
Kevin Edwards, Fantasia, Dallas, Texas
Donald J. Embree Antiques, Inc., Dallas, Texas
Puddin Evans, Mondays Antiques, Dallas, Texas
David Harris Antiques, Big D Bazaar, Dallas, Texas
Jean's Antiques and Collectables, Houston, Texas
Elizabeth Leconey, Highland Park Antiques and Nauticals Ltd.,
    Dallas, Texas
Esther Maldonado, Ester's Antiques, Bandera, Texas
Ward and Don Mayborn, Uncommon Market, Dallas, Texas
LaDonna Mechaley, Serendipity Shop, Rapid City, South Dakota
Barbara and Thomas Morrison, The Victory Antiques, Dallas,
    Texas
Motif International, Dallas, Texas
Rosie O'Reilly Antiques, Austin, Texas
Our Favorite Things, Dallas, Texas
Judith Peters, Big D Bazaar, Dallas, Texas
Nadine Reynolds' Antiques, Austin, Texas
Return Engagement, Dallas, Texas
Dr. Harvey Richman, Le Monde Antiques, Dallas, Texas
Kathleen M. Russell, Big D Bazaar, Dallas, Texas
The Sandbergs, New Braunfels Antiques, New Braunfels, Texas
Bud and Betty Sparks, The Attic, Bryan, Texas
Doris Turner's Antiques, Austin, Texas
Joe Webb and Jerry LaFevor, Paradise Antiques, Dallas, Texas

The cover photograph is an Air Raid Bell of the type used in England during World War II. It is 10½"h and quite similar to a School Bell. Courtesy of the Uncommon Market.

Preface

The world of brass is a fascinating topic of study. The metal, although known in prehistoric times, has continued throughout the centuries to the present day to benefit civilization in countless ways, both functional and decorative. The metal's resemblance to gold has perhaps sustained and maintained interest in brass throughout history. That is not to say that the metal has not had its critics or ups and downs in popularity. At times "brass" has had a bad connotation and been the victim of low repute, especially when the metal has been used falsely to represent gold. We have all seen old movies where someone bites a coin to be sure that the coin is in fact gold and not brass! "Brassy" is a word still used which has a degrading meaning, defined by Webster as "cheap and showy." The origin of that term appears to have come about during the late Victorian times when an over abundance of brass was used by the lower classes for decoration and ornamentation!

It is apparent that brass has managed to endure and survive its critics, however. Today, tremendous amounts of new brass are imported to the United States, showing that there is a profitable market for new brass goods of all types. But most importantly, there is a very large market for old brass today because of its scarcity and the past it represents. New brass may strive to emulate and copy old brass, but there is a difference between the two, making the older pieces highly sought.

This book is designed to show what is currently available, basically in the realm of European and American brass, at various antique retail outlets such as shops and shows. Most of the items featured are products of the late 18th century through the mid 20th century. Although brass made during the first half of the 20th century is certainly not "antique," brass of that period is now beginning to find its place on today's antique market. Many of the items from the early years of this century are obsolete in use and thus collectible for that reason. Other brass reflects the styles of the Art Nouveau and Art Deco periods of this century and merits collecting because of the distinctive styles associated with those periods. Even decorative brass items such as plaques and jardinieres featuring embossed tavern scenes and fruit themes which were popular during the 1950s are now considered collectibles.

Brass made prior to or during the 17th century and early 18th century in Europe and America is quite rare on the open market. The surviving examples from those eras have been acquired in most instances by museums or private collections. While such museum pieces are definitely important and useful in tracing the history of brass, brass from those sources has not been featured here because those items do not come within the scope of this book--to show what is currently available for sale in old brass today.

The book includes a brief historical survey of the subject and some rudimentary facts about brass-making are discussed and defined. Tips on collecting brass and ways of differentiating between old brass and the new "repros" are also included. Over 300 color photographs of brass objects have been divided into six broad categories. An Object Index is also provided to aid in locating specific items. Because brass spans so many collecting areas and because numerous brass items are unique or one-of-kind, it is not possible to indicate an average price range for all brass objects. Thus prices reflect the dealers' "asking" prices. Except for a few pieces, all brass was for sale at antique shops and shows. Those few examples owned by individuals were priced on information derived from research and comparison to similar items for sale on the retail antique market.

The majority of old brass available today was made during the machine age, but enough different brass items have been made to appeal to a varied group of collectors. This interest ranges from the most specialized areas of collecting such as match safes, bells, and thimbles to more general collecting fields including kitchen collectibles, nautical antiques, and railroadiana to name a few. Moreover, brass is increasingly attracting many "non-collectors" who desire old brass to add decoration and style or authenticity to their home or work place. Old brass has indeed become a coveted acquisition to a broad cross-section of the American public. Current prices reflect that demand and supply.

By taking a look at brass manufactured over a time period of 200 plus years, we admit that the door has only been "cracked" so to speak in providing a "peek" into the wide world of antique and collectible brass! Hopefully, this book will serve as a useful and practical guide to that subject for both beginning and advanced collectors.

Mary Frank Gaston
P.O. Box 342
Bryan, Texas 77806

Readers wishing to correspond, please include a self-addressed, stamped envelope if a reply is requested.

Brass and Its Production

Brass is a metal alloy made from two natural elements, copper and zinc. Copper constitues the bulk of the formula for making brass, ranging from about 66 percent to 83 percent. The amount of zinc added to the copper affects the resulting color of the metal and also adds strength and durability to the alloy. The optimum color for brass is considered to be a rich gold. A gold color is obtained when the zinc content ranges between 17 and 30 percent. Zinc used in proportions of over 30 percent gives the metal different colors varying from shades of white to gray. The less zinc used, the redder the color of the brass.

Brass has certain properties which make it a very useful metal. It is quite flexible and lends itself to being shaped in many different ways. Brass is harder and more durable than copper alone. Brass is not rigid and does not break easily. It is a good conductor of heat and does not rust. The alloy can also be used as a base metal for other finishes such as silver. The finish gives the look of the "real thing," but at a fraction of the cost of an item made entirely of a more precious metal, such as silver or gold.

The chief disadvantage of brass is that it should be lined with some other metal if the article is to come into contact with food. Most foods have a "tainted" taste if served from unlined brass vessels. Acidic foods can cause a reaction with the metal on direct contact and corrode the brass itself. In the early days, people were not knowledgeable of these effects. In time, they tried to correct the situation by lining brass (and copper) utensils with tin. There was still a problem though, because lead was required for the process and it was poisonous. By 1756, in England, however, the technique of tinning brass and copper without using lead was perfected (Wills, 1968, p.21). It should be pointed out that some countries still did not line all brass cooking vessels even after it was possible to use the pure tinning process. Milk containers in Holland and apple butter and candy-making kettles in America are noted examples. Eventually nickel and then stainless steel replaced tin as a lining for brass and copper cooking utensils.

A second slight disadvantage of brass is that, over time, it tarnishes from exposure to air. Thus, if one desires brass to be shiny, like gold, the items must be polished. This is easily accomplished with the aid of commercial polishes although it does require some time. Alternatively, brass can be lacquered which will deter tarnishing and only necessitate dusting from time to time.

From the multitude of items which have been made and continue to be made from brass, it is apparent that the metal has well withstood these two disadvantages. The metal's wide diversity has enabled it to remain a very popular metal for manufacturing a wide number of both utilitarian and decorative items for hundreds of years.

Origins of Brass

Bronze was the predecessor of brass. Bronze is considered to be the oldest known metal alloy. Its discovey occurred in Mesopotamia centuries before Christ. The "Bronze Age" takes its name from this alloy and marks the second period in the cultural evolution through prehistoric times, succeeding the Stone Age and preceding the Iron Age.

Brass is not as old as bronze, although the actual discovery of brass (who, when, and where) is not very clear. The reason for this is because bronze and brass often were confused in ancient times as being one and the same, and sometimes they are still confused. The two alloys have the same basic ingredient: copper. But tin, rather than zinc, is combined with copper to make bronze. In most instances, the color of the two alloys is different. Bronze is usually reddish-brown while brass is gold. But when less zinc was added to brass, the color was reddish and thus similar to bronze in color. When either of the two metals was gilded, it was again often impossible to distinguish between the two. Even today, bronze and antique cast brass or gilded pieces are sometimes identified as the wrong alloy. A chemical analysis would be needed in most instances to determine whether the item is brass or bronze.

Brass was actually superior to bronze in several respects. Bronze was not as easy to work with as brass. Nearly all bronze items had to be cast into shape. Brass could also be cast, but its ductile nature allowed it to be fashioned in other ways from the earliest of times. The bronze alloy was easier to achieve than the brass alloy, but the workable nature of brass caused brass to surpass bronze in use through time.

Evidence confirms the fact that brass (as differentiated from bronze) has been found in remains of ruins in various parts of Europe and Asia predating modern civilization. The Roman period (27 B.C. to A.D. 395) is considered by some to be the earliest time when brass was made. Through most of the Dark or Middle Ages in Europe (A.D. 476 to 1450), the art of making brass and bronze was lost. The processes were revived to a great extent, however, by the 13th century. One exception to this is the town of Dinant, in Belgium. That city had a thriving brass industry for over 500 years. In 1466, the entire city along with its brass industry was destroyed. Aachen and then Nuremberg in Germany were also famous brass producers from the 15th through the 17th centuries. Central Europe, especially the Germanic areas and the low countries such as Belgium and Holland were well suited for the development of mineral industries. Vast mineral deposits and rivers in those regions made the locations natural for the establishment of such industries. The ore which contained zinc, and of course necessary for the manufacture of brass, was especially plentiful. The proximity of these countries to each other caused the methods and techniques of making brass to be carried from place to place and country to country. By the

1600s, brass was firmly established as an important trade commodity throughout Europe and other parts of the world.

The similarity of brass to gold, both in color and flexible nature, accounts for the development and refinement of the brass-making process from Medieval times to the present. Gold has always been a much-coveted mineral. Historically, attempts have been made to make gold from other materials rather than leave the occurrence and discovery of the precious metal to the whim of nature and chance. During the Middle Ages, alchemists tried to find a formula for gold by mixing together various natural elements. Alchemy was a very secretive and mysterious process. Brass-making during this period of history was identified with the alchemists' work. The process of making brass remained cloaked in secrecy for many years. As with all secret processes, the technique was eventually passed on from one person to another, from one country to another, or discovered independently. By the 17th century, the method was being recorded and published. Finally, after many centuries, the secrecy, mystic, and magic surrounding making metals from alloys was lifted. The processes gained acceptance and recognition as a branch of true science called metallurgy.

The Manufacture of Brass

From the Middle Ages until near the end of the 18th century, brass-making was not an easy undertaking. The methods of extracting the minerals from their ores were crude, and the by-products given off by the melting of the ores were poisonous and thus hazardous for the workers. Although bronze was always made by directly fusing copper and tin, it was not possible to make brass by direct fusion in Europe until 1781. (As with other Oriental advances, China had been able to manufacture brass by direct fusion centuries earlier.)

From the Middle Ages until 1781, European brass was made by crushing the mineral ore *lapis calaminearis* to mix with the copper ore. Haedeke (1969, p.28) notes that the Europeans did not even know that zinc was the actual element in the ore. They only knew that the crushed ore when mixed with the copper ore, would change the color of the copper. The process of crushing the zinc ore was involved and complicated. For that reason, brass centers were often located near the deposits of the *lapis calaminearis* rather than the copper sites. The copper ore was easier to obtain and transport than the zinc.

Once the two ores were properly crushed, they were heated at high temperatures until they became molten and blended. After the mixture had cooled somewhat, it could be poured into pits or onto slabs of stone. When the mixture was poured onto slabs of stone, it formed a sheet which could then be cut into strips, or the sheets could be hammered into desired thickness. The brass which was poured into pits was used for casting objects.

In that time period, without benefit of advanced technology, brass-making was often a "hit or miss" activity. It was not possible to control the measurements of the ingredients or the temperatures for melting the ores. As a result, the quality, texture, and color of the brass often would vary greatly. Great skill, time, and effort were necessary to produce brass. Nonetheless, from the rather crude and inexact techniques of the Middle Ages, many brass articles were made in this manner up through most of the 18th century. But with practice came expertise and improvements. Such experimentation leading to improved methods of manufacture was warranted because of the continued demand for brass items of all types. The metal was functional, and its resemblance to gold helped to brighten up the mostly dark and drab homes of the middle classes.

Early methods of shaping brass were beating or hammering the cooled molten metal. This was done by hand in the early days and later by water powered machinery. Objects were literally "beaten" into the desired shape such as a kettle or a pan. "Battery Works" became the name coined for locations where brass was made. From an early time, brass was also drawn into wire and made into pins. The wire was used to make the teeth in wool-cards. Such cards were essential for obtaining yarn from the raw wool. Brass pins, of course, had many uses and were a great improvement over inferior wire that had been used in the past.

There were several ways of casting brass objects. One was called the lost wax (*cire perdue*) method. That technique was replaced by sand casting during the 1700s. (See the section on "Useful Terms" for explanations of those processes.) Brass could be cast solidly, that is with no hollow center, and much early brass was cast in that manner. Later brass was cast in two parts and soldered together. Core-casting was an improvement over two-part casting because the article could be cast in one piece with a hollow core.

As well as being a relatively easy metal to shape, brass also could be decorated with various designs applied in a number of ways. Chased and engraved designs, pierced patterns, and repousse' work were popular methods, accomplished by hand at first and later by machines.

The English Brass Industry

Although brass was and is made in many parts of the world, the English Brass industry had the greatest impact on the development of brass-making after the 17th century. England was late, in relation to the other European countries, in becoming involved in making brass. During the 18th century, however, the country became famous the world over for the finest brass made. Many improvements in manufacturing brass were developed and patented in England during the 18th century, and thus it is important to take special note of England's part in any discussion concerning the history of brass. England also had a direct influence on the development of brass-making in America. Moreover, many items which are available for collectors today are of English origin which further makes the history of English brass pertinent to American collectors.

The detailed history of England's brass industry is quite fascinating and has been well documented by several English authors. Some titles in the bibliography are recommended for those interested in reading about that somewhat colorful part of English history. For our purposes, we will limit the discussion to several facts which are considered basic knowledge for brass collectors.

Until the latter part of the 16th century, England relied on imported brass from other European countries, especially Holland, to supply her needs. That situation began to change when Queen Elizabeth I of England began her reign in 1558. The Queen was responsible for opening the way for England's entry into the brass industry. She realized that England must become independent in as many ways as possible in order to remain a sovereign power. Most importantly, it was necessary for the country to be able to make its own weapons and tools of defense. For armaments, metals were needed. The first step toward such independence was to bring all of the mineral mines in England under royal ownership. That was accomplished in 1568. The Society of Mineral and Battery Works was also established at that time to manufacture articles from metals, especially arms. The royal monopoly of the mines and manufacturing lasted almost one hundred years, until England's Civil War with Scotland in 1642. That war brought about the collapse of England's brass industry as well as the destruction of most of the available brass objects which were melted down to make more arms (Gentle and Feild, 1975).

The Civil War ended in 1649, but the British brass industry did not get back on its feet for almost forty years. In 1688, William of Orange (William III) became England's first consititutional monarch. In the next year, the Mines Royal Act was passed which took control of the mines out of the hands of royalty. Thus it became possible for individual ownership of mines and metal manufacturing. Consequently, brass-making in England began once more in earnest.

During the 18th century, many towns in England manufactured brass, but Birmingham became the best known center of the English brass industry. Throughout that century, England made and exported a tremendous amount of brass of all types. As early as 1699, England prohibited her American colonies from performing any type of manufcturing, including that of brass, thus assuring a large and constant market for her products in that part of the world. But English brass gained renown abroad as well.

During the 1700s, events occurred in the English brass industry which revolutionized the manufacturing process and eventually brought about its total industrialization. The two most important events were the discovery of distilling pure zinc in 1738 by William Champion and the discovery of making brass by the direct fusion of copper and zinc in 1781 by James Emerson. Distilling pure zinc eliminated the tedious process of obtaining the metal from crushing the *lapis calaminearis* stone. Distillation made it possible to have pure zinc in the form of ingots. The direct fusion technique (43 years later) was the real breakthrough for the industry. That technique enabled the manufacturing process to be controlled which resulted in a standard quality and color of brass. The best English brass was said to have a composition of one quarter zinc to three quarters copper. That formula resulted in the best gold color.

Several other important patents were taken out in England during the 1700s which also played a decisive role in the industrialization of the brass industry. A patent for rolling brass by machines was a vast improvement over beating the metal into sheets and shapes. A stamp and die method of placing patterns and designs on brass was patented by John Pickering in 1769. That invention replaced the necessity of decorating brass totally by hand. In 1777, John Marston was able to stamp out small items entirely by using the basic principle developed by Pickering. Eventually, it was possible to stamp out larger items, and soon stamping largely replaced casting as a method of shaping brass objects. The invention of steam power by James Watt in 1769 was adapted to brass-making machinery during the 1780s. By the beginning of the 1800s, Britain's brass industry was definitely industralized.

Mechanization of the brass industry meant that a large number of different items could be produced quickly, cheaply, and efficiently. Mass production was good on the one hand because it allowed production to keep pace with demand and because more people could afford the products. On the other hand, such an abundance of brass at lower prices was responsible for the metal losing its appeal among the middle classes. Brass became cheapened in the eyes of many and began to be considered less desirable. Although the machine-made brass produced during the early 1800s retained the fine qualities aassociated with earlier English brass, and in spite of the fact that earlier brass was considered superior to those products made by machines, many fine brass pieces from those earlier times were thrown out, melted down, or stored away to be out of sight and forgotten during the 1800s (see Gentle and Feild, 1975, p. 57). Luckily some of those pieces were saved and salvaged by astute collectors in later years.

English brass made prior to 1800, before total mechanization of the industry had arrived, remains scarce. Many examples which have been discovered are now in museums today. Occasionally a few of those early items surface on today's market. Things that go out of style have a way of coming back in vogue after a period of time, often on a higher level of appreciation (and hence price) than when they were first made—Tiffany Lamps are a good American example of that type of trend. Such examples of English 18th century brass are of course highly desirable when found, and if one can pay the price.

Many historians and antiquarians consider that only brass made in England before about 1850 should be classified as truly antique brass and merit consideration for a collection. The era of machine-made antiques and collectibles has definitely arrived, however. Furniture, glass, and all types of ceramics are just a few categories where this is true, especially among American collectors. Machine-made brass is another one of those areas.

Because of a lack of supply of earlier items, those who appreciate the metal for its aesthetic qualities turn to English brass made during the late 1800s through the early and even up to the mid 1900s. Many of those articles are over one hundred years old or close to that age while others are definitely "collectible." Brass from that period is relatively plentiful in supply. Some prices are even quite high, but overall a sufficient quantity of brass of that vintage is available which not only spans a broad range of collector interest, but also spans a broad range of pocketbooks!

The American Brass Industry

Earlier, I noted that England prohibited America from carrying on any type of manufacturing after 1699 (and until about 1776). Brass had been made in the colonies on a limited scale prior to that time, and the new law did not really put an end to the brass-making that was being done. The law was a difficult one to enforce. Most of the people in America who were making brass did so as a "cottage-type" industry. At that time the colonists did not have access to the raw materials for making brass. The brass made was by people called "braziers" who used scrap brass, bits and pieces of old or worn out imported brass. The scrap metal was melted down and new items formed. The braziers often traveled around from place to place plying their trade.

Most American brass of that period was of a utilitarian, rather than decorative nature. Its uses were mainly for lighting, heating, cooking, and washing--the basic necessities of life. Any item made of brass was important for any household. Brass was highly esteemed in the early American home and remained popular for a much longer time than it did in England. In fact, brass possessions were an indication of prestige and owned by the wealthy in the colonies whereas in England, the metal was owned mainly by the middle classes. All brass was treated with great care and passed on from one generation to another. Many examples show signs of careful mending.

It is generally agreed that brass-making on more than just a modest scale did not gain any great momentum until after the Revolutionary

War. Even after the war, when the colonists were no longer forbidden by "law" to engage in that type of industry, it was still not easy to make brass from "scratch." The minerals had to be imported, especially zinc, and native copper was not mined to any great extent during that period. As a result, much American brass continued to made from scrap. As more and more people emigrated to American from England and Europe, many brought advanced knowledge and skill in making the metal itself and thus helped to widen the industry in this country. Eventually some brass-making was carried on in most colonies. Pennsylvania and Connecticut were among the most prolific. All types of brass were manufactured. Brass buttons were very popular, and of course, all kinds of kettles and cooking utensils were made. Early American brass was made by sand casting or by hammering sheet brass. A new method of shaping brass was introduced in the 1850s by H.W. Hayden. He developed a technique for spinning brass which made the articles much lighter in weight which was especially important for large cooking kettles and pans. This particular method of making brass was practiced to a large extent in Connecticut, and many of the pieces were signed by the makers (Ketchum, 1980, pp. 142-143). Signed brass from this period is quite rare because brass made in Europe or America was seldom marked prior to the late 19th and early 20th centuries.

Although America was free to make brass after 1776, and the industry did grow, a large amount of foreign-made brass, including English brass, was still imported. Historically, that situation has not changed because foreign brass has always been less expensive to produce than American brass, and thus the price has been cheaper. The industry in this country has not been able to compete too profitably with the imported products. Therefore American-made brass is relatively scarce in comparison with brass made in other countries.

Collecting Brass

This book focuses primarily on European and American brass rather than brass of other origins. Collectible or "old" brass is considered to be any brass product made through the mid twentieth century, ca. the 1950s. Brass made prior to the twentieth century is of course more desirable, but such examples are quite scarce today, and when available are quite expensive. Many items made in brass during the early part of this century merit collector status because the piece is unique or obsolete. Brass made as late as the second quarter of this century is now more than thirty years old and thus considered collectible. Examples are more plentiful from this later period and are usually fairly easy to distinguish from current imports and the "repros" which are on the market today.

Objects made of brass have been collected as a source of pleasure, curiosity about the past, a link with history, and as an investment by serious collectors for many years. Interest in brass today, however, spans many collecting areas and is not confined to brass collectors per se. Brass plays an important part in general metal collections which may also include copper, tin, and pewter. Many collectors especially like to concentrate on copper and brass for a collection. One reason for this is because many articles were made by combining the two metals. A dipper or bedwarmer may have a copper bowl but a brass handle, or sometimes copper and brass were used to achieve a decorative effect. The two metals complement each other and make attractive displays.

Collectors of tools and scientific instruments compete in the search for brass. Navigational equipment contains many items made of brass, and many collectors are interested in the subject of nautical antiques. Kitchen collectibles of all kinds are rapidly gaining interest among a large segment of the collecting public. Brass, of course, finds a place in that particular area.

But most importantly, old brass has broken out of the bounds of antique collecting into even more general demand by persons who might not be termed "antiquers" or "collectors" at all. The concern and interest in home decoration has been responsible in a large way for bringing antique brass to the high status of attraction it is currently enjoying. Interior design and decoration has become a big business in the United States. Many books and magazines are totally devoted to illustrating ways to make the American home attractive as well as functional. Decorating firms are located in cities and towns all over the country. Brass is recognized as a metal which can bring warmth, charm, and style into a home of any size. Brass accents for totally decorative purposes fit well into the most modern homes and apartments. Most of the brass pieces in such instances take on a use different from what they were originally intended. Coal scuttles hold dried floral and leaf arrangements; cooking kettles and pans may hang on a kitchen wall but are not for use--only looks; and yacht tie-downs and sextants in wooden cases may serve as paper weights and bookends respectively. While new brass items can and may fit the need or achieve the look just as well and perhaps at a fraction of the cost, many people prefer the old, and are willing to pay top prices for just the right object.

Older homes are regaining popularity on the home-buyers market. New owners often wish to restore the homes to their original period as much as possible. Authentic brass hardware and lighting fixtures are in demand for such purposes where they serve functional needs in addition to being decorative and "period." Over the last few years, there has been a mounting interest and trend in building new homes in Victorian styles. Thus another market opens for brass fixtures and accessories. There are even those individuals who have opted for getting back to basics in earnest, especially to conserve energy and fight inflation. They really use many of the earlier methods of lighting, heating, and cooking thus widening the demand for old brass in the way of kerosene lamps and lanterns, fireplace tools, footmen and trivets, and kettles and pails of all types. A very wide umbrella does indeed cover the many interests in brass today!

Sources and Availability

Because of the ever growing interest in brass of all types coupled a few years ago with the increase in the price of the base metal, copper, old brass is becoming harder to find and prices are high and getting higher. Newcomers to brass collecting might wonder just how much old brass is presently available today, and where the sources are for such pieces.

Early American brass is much more scarce than brass of other origins. Examples which definitely can be authenticated as American are mostly in museums today or in the hands of advanced collectors and not available on the open market. The United States still remains a hunting ground and source of old brass, however. Collectors can scour small shops and out-of-the-way places and find genuinely old items which have been in this country (even if perhaps not made here) for a hundred years or more. They will be more successful, of course, in finding brass made during the late 19th and early 20th centuries.

The largest supply and source of old brass on the American market is still England. For over ten years, hugh cartons of English antiques of various types have been imported into this country. Whereas many dealers historically have gone abroad to select and bring back European antiques for American customers, the large shipments came about in response to the greatly accelerated interest in antiques by the general public over the last fifteen or so years. Many brass items arrive in these lots. Some of the pieces are definitely old which have been salvaged from Britain's 18th and early 19th century brass industry. Others are objects made as late as the 1940s and 1950s. Most of the pieces are in fact English, but some are of other European origin. Similar containers are also being imported from France, Holland, Portugal, and Spain. Such shipments are sold to antique dealers throughout the United States. More and more, large shops are relying on this source of supply to meet the needs of their customers. A variety of interesting brass items will usually be found at such locations.

Metal stripping shops are another source of old brass. Many metal items such as brass and copper were plated with another metal at one time. Today the underlying metal is much more valuable and attractive than the plated items which have begun to show signs of wear and are relatively expensive to re-plate. Special shops focus on stripping such ware as well as polishing and lacquering the pieces in the process. Often these businesses operate as retail outlets as well as performing the service for individually owned items. Trays, kettles, fire extinguishers, and all types of hardware are some examples which may be found for sale.

In the same way, to some extent, brass hardware and lighting fixtures can be found in businesses specializing in "architectural antiques." All types of fixtures have been rescued from older homes in this country which were about to be demolished for "progress." Windows, flooring, mantles, columns, ceiling fans, and sometimes whole rooms can be located at these stores in addition to hardware and lighting fixtures. The prices are often quite competitive with new fixtures. Many businesses today specialize in making new hardware which looks old and can be used to complete old items lacking drawer pulls, key holes, and knobs, but if one desires, authentic old pieces for those purposes can still be found.

Old and New Brass

The early brass makers did not furnish precise clues to enable collectors years later to attribute a specific brass object to a specific maker or even to a specific origin. English brass of the 17th and 18th century was not required to be marked even though English silver was. Similarly, American makers also rarely marked their brass. During the latter part of the 19th century, and on through the 20th century, brass was marked more often by the manufacturer because of the stamping process. "England" or "Made in England" is often seen on later English brass, even if there is not a name of a particular manufacturer. Thus, it is indeed a "find" if a marked exmple of English or American brass is found made prior to the late 1800s or early 1900s. Collectors are advised that a name on a piece of brass does not necessarily mean that the item was made by that person. Many people had their name engraved or placed on their brass possessions to indicate ownership. Furthermore, a name without a specific location, such as a town or a country, cannot be used definitely to identify origin of the object.

Styles or shapes of brass also cannot be used conclusively for dating brass. English brass imitated earlier European styles and American brass copied English designs. Because English brass often was made along the lines of English silver, hallmarked silver items have been used as guidelines for dating early English brass. That method is not foolproof. Even though a brass object may look just like a dated silver item in shape and design, that does not mean that the brass was made at the same time. It probably was made at least somewhat later, though it may be styled along the lines popular during a certain period-- Georgian or Queen Anne for example. Although brass copied the silver styles, brass itself from those periods also has been copied. Today many of the "copies" are now quite old and are collected appropriately as old brass.

During the 18th century, the English had pattern books illustrating brass items. These books were similar to what we would call catalogs, and were used by salesmen to show the type and variety of brass items available and their prices. The catalogs were rarely dated, but they illustrate what was being made in brass during that century even though exact years cannot be pinpointed. Old paintings showing brass objects have been suggested as another means of dating brass. Such examples may convey that certain items and styles in brass were being made during the time the piece of art was painted but such evidence is not conclusive. The same is true when comparing an item to one found in museums. Unless the item is marked just like the one featured in a museum, the same style cannot definitely identify the item as one by the same maker or from the same time period or same origin.

Type of item is also not a clue to the age or origin of brass. Most objects of brass have been made for years, copied again and again by European and American brass makers. A few items however are noted to be especially indigenous to a particular area. Objects with hinged lids are thought to be European and not American. Andirons were more prevalent in America than in England because from a very early time England heated with coal rather than wood. Samovars are considered Russian, Turkish or Far Eastern in origin rather than West European. Apple butter kettles appear to be an American invention.

New brass from all over the world continues to flood the American market today. There are stores which sell only new brass. All furniture, gift, jewelry, and discount houses, as well as variety stores, stock a large selection of new brass. These new items generally do not pretend to be old or antique although they may be fashioned along traditional lines. Some importers, however, have items made that definitely replicate what collectors consider as "old" brass objects. Candleholders, tea kettles, jelly kettles, trivets, coal scuttles, school bells and so forth are some examples. Among collectors, such items are termed "repros" because the articles have made to resemble genuinely old items which are being sold on the antique market. In most instances, the "repros" are sold at many of the same locations when authentic antiques also are sold such as flea markets, antique shops and antique malls. The new pieces are solid brass. They may have a paper label saying where they were made (often Taiwan). The labels do not remain on for long, and thus to the unsuspecting buyer, the item may be thought to be old.

In some ways the great influx of new brass and especially the "repros" have been somewhat detrimental to old brass collecting. As in Victorian times in England, there is so much new brass available that those searching for the old sometimes begin to feel that the old has become somewhat cheapened in the process. But from my personal observations and discussions with collectors and dealers, it is apparent that there is really never a problem in selling old brass. A large segment of the collecting public is still determined to turn their backs on the new and the "repros" and diligently search for the older pieces.

Tips on Identifying Old Brass

If old brass cannot be dated easily because it is not marked and because decorative styles and types of items also cannot be definitive as dating guidelines, just how can one differentiate old brass from new brass? From talking with brass collectors and dealers, it is apparent that not being able to date a brass object specifically does not hinder the collector. Brass collectors are not rigid in demanding that brass must be signed, marked or noted as having been made at an exact time. Collections would be very limited and not affordable for most collectors if that were the case. By an understanding of the past through reading books, visiting museums, discussions with others with similar interests, and by being aware and alert to the new brass, the collector becomes aware of certain clues that help in distinguishing the old from the new. These methods are the most thorough way and best protection against getting "stung" in any area of collecting.

Some suggestions or "rules of thumb" can be used to help distinguish between new brass or "repros" and brass made prior to the mid 20th century. The weight of the brass is a good starting point. Old brass, including late 19th century and early 20th century brass, is heavier than brass currently manufactured. New brass is thinner and sharper because the brass is rolled much more thinly. New cast brass is also usually lighter is weight. Compare new cast brass hardware to older pieces, for example.

Examine the object in detail. Look for signs of wear and smooth edges on old brass. Inspect the bails or handles on kettles and pails. These were usually made of iron and rounded, although some were also made of brass. New bails on the "repros" are often just a thin strip of flat metal. See how the piece is constructed. Was it made in one piece, are the sides seamed, is the bottom dove-tailed? In older pieces, the soldering metal may show slightly in the seams, but beware of items where the soldering is too obvious. Soldering is usually noticeable in dovetail construction, but that method of manufacture is too time-consuming and costly today.

Signs of mending such as on the bottom of pans and kettles indicate that pieces are old. Because many cooking utensils were lined with tin, evidence of the tinning should remain (unless the piece has been stripped). Brass is also sturdy and does not damage easily. If the piece is dented, try to determine if the dent occurred in the past or has the object been "newly" dented to look old.

Be aware that some brass pieces have been painted over the surface. I once found a beautiful double student lamp which had been covered with gold spray paint (not gilded!). I imagine the piece had tarnished badly, and the paint was the solution to the discoloration and a remedy for polishing! But from the style of the lamp, I thought the piece should be brass, and when my magnet did not attract at any point, I bought it. Sure enough, the paint came off with paint remover, and after being cleaned and polished, I had a beautiful brass lamp!

Brass that has not been polished acquires a "patina" over time eventually turning the metal a dark color. Sometimes this patina is a useful guide in telling whether brass is old or new. The patina is often so dark that the piece may look more like bronze, or it may look like a cheaper metal alloy. Don't overlook such pieces which may not readily catch your eye because they are not shiny and gold like brass. New brass also has a tendency to turn red or pinkish in color as it tarnishes rather than acquiring the dark patina of old brass.

Most new brass is lacquered today to keep the metal from tarnishing, but lacquered examples do not mean that the item is new. Some people do not like the dark patina. They prefer the rich gold look plus the fact that lacquered pieces do not have to be polished. Dealers tell me that although it is a added expense and thus increases the price of an item, most customers prefer to purchase old brass which has already been lacquered. Lacquering brass is not a new technique, but has been used for hundreds of years. Old brass which was lacquered many years ago, however, will by now have signs of the lacquer wearing off if it has not worn off already.

Always remember to take a magnet along when searching for old brass. Some items are merely brass plated. If the magnet does not attract, the piece is solid brass. Some items, such as lamps, however may have an iron bar through the center which will attract the magnet. In such instances, one should check out the other parts of the lamp as well.

Price is often another indicator in determining whether brass is old or new. Most new brass is relatively cheap when compared with prices for old brass. It is still possible to find some real bargains in old brass today, however. Some new finely made brass is also quite high in price. Examples of the latter do not usually cause confusion with the old because they are offered for sale by department stores or specialty catalogs where it is obvious that new merchandise is sold.

The "repros" that are so much in evidence at various "antique" locations are decidely cheaper than the authentic item. Although over the past few years the prices have been increasing. I suppose that is in response to the fact that genuinely old obejcts are becoming more expensive. Once in a while, a "repro" will have a price tag equal to that of its older counterpart. I am sure some people purchase such pieces thinking they are getting a true antique or collectible. But if a school bell is $15.00 or a jelly kettle $35.00 or a pair of candleholders $17.50, it is usually safe to assume that the objects are new.

Be suspicious when you see a whole lot of identical objects such as coal scuttles, jelly kettles, trivets, or ships' lanterns in the same place or identical items in gift shops, flea markets, or antique malls. The "delft-type" handle on some brass is an instant clue that the piece is new.

Do not hesitate to ask questions of dealers. Most are quite willing and happy to tell you what they know about an object such as where they bought the piece, what it was used for, and what period of time they think that it was made. Shops importing from England or other parts of Europe often have very good background information concerning those articles. Dealers who carry new or "gift" items as well as antiques will usually tell you which pieces are new or old if you ask. Sometimes they will often volunteer the information when they notice you are interested in a particular object. Make friends with the dealer, and you will benefit from the dealer's knowledge and expertise.

Please note that most of the brass illustrated in this book was made during either the latter part of the 19th century or early part of the 20th century. Therefore dates are not shown in the captions for most photographs. If the pieces were made at other times, the captions may indicate other time periods such as 18th century, early to mid 1800s, or ca. mid 20th century (1925-1950). Remember it is usually possible to date brass only according to approximate time periods.

Major Categories of Collectible Brass

Of the multitude of objects that have been made of brass over the past several centuries, it would be impossible indeed to show or discuss examples of each one. Several hundred photographs representing a wide range of brass items are featured in this book to give a broad view of what is currently available on today's market. I have used six categories (listed below) for the brass presented in this book. Obviously some items may fit into more than one category. Many of the items in a particular category could easily constitute a book alone--candle holders and andirons, for example. An Object Index is provided at the end of the book so that readers may look there for specific items if the object in question is not immediately visible in one of the categories, or to see if a particular item has been included in the book. Although I have tried to show a representative sample of brass items in each category, I certainly do not claim to have completely covered this vast subject!

Note that some of the items shown are made only partly of brass. Brass items were combined with other metals and even wood. Brass was used in this way because it was necessary to the function of the piece, or its use was to enhance the beauty and exhibit signs of fine workmanship on the object.

1. *Lighting Implements.* Brass lighting implements are one of the most popular areas of brass collecting. They are both useful and decorative. Candle holders, candelabra, wall sconces, kerosene lamps, and table lamps are included. Chandeliers are not featured due to the breadth of that subject alone, but collectors should be aware that a wide variety of beautiful brass chandeliers is available, especially at businesses specializing in architectural antiques and stripping shops. Prices are quite competitive with new chandeliers.

2. *Fireplace Accessories.* A look at objects in this category probably makes us grateful for central heating and the ability to enjoy bits of the past for decor without really having to "use" everything that was once connected with fires and the hearths in the past. Firedogs, andirons, fenders, fireplace tools, coal buckets, wood boxes, screens, footmen and trivets are some of the items featured. Remember the fireplace in the kitchen was the center of activity in most homes. Its use was not only for giving warmth, but during certain periods it was used also for light and cooking as well. Brass used for cooking purposes at the fireplace, however, is included in the following section on Kitchen Collectibles.

3. *Kitchen Collectibles.* Large kettles and pans for cooking, and smaller kitchen tools were often made of brass or partly of brass. It appears that more utilitarian kitchen items were made of copper rather than brass, however. Serving pieces which would probably have been found in the dining room such as coffee services, teapots and burners, and trays are shown in this section also. Some of these latter items were once silver plated.

4. *Tools and Instruments.* Brass was well suited for certain tools and scientific instruments because of its strength and non-rusting qualities. Various trade tools and instruments are shown. A fairly large section of nautical items is also presented. A few examples of furniture made from some of these tools are also included.

5. *Hardware Fixtures.* Practically any type of hardware for all

puposes was made in brass. Door bells, mailboxes, furniture mounts (ormolu), curtain tie-backs, drawer pulls, and locks are just a few. Old brass hardware is, of course, functional, but interesting examples can also make attractive accent pieces or form a unique display.

6. *Decorative and Personal Objects*. This category covers a very broad spectrum and could be expanded indefinitely. Many of the pieces illustrated were originally designed to be functional as well as decorative. Mirrors (frames), desk and writing items, smoking accessories, and even bird cages are such examples. Plaques, jardinieres and vases have always served a decorative purpose. Some personal items such as the bidet were once totally functional, but now purely decorative, I'm sure!

Prices

A price is quoted for each brass item illustrated in this book. With few exceptions, the prices are those that were taken at antique shops and shows. A few examples which came from private collectors have prices based on what dealers were asking for comparable items. The price information in this book should be used only as a guide. It is not intended to set prices. The prices represent a random sample of current prices being asked for brass today. You will notice, however, that many of the similar items, although they came from different sources, are close in price, and thus do in fact suggest a range of price for those items. Because brass covers such a broad area in many respects including, age, origin and article, many examples on today's market have similarities, but they also have certain differences which may cause one item to be much lower or higher than the other in price.

Several factors always play a part in the final purchasing price of any antique or collectible. The overall condition of the item is important. Whether it has been mended, has new parts, or is missing a part are important considerations. For brass, not only those factors, but also whether the piece has been lacquered or stripped and lacquered, or if it has been made into a lamp or a piece of furniture will add to the price. Older items and unusual pieces are expected to bring higher prices than more common-type or later pieces. Signed examples from early periods command a premium, and American-made brass is usually favored in this country, and thus may be higher in price than brass of other origins. Additionally, prices are higher or lower in various regions of the country and may vary even within the same area. Sometimes articles at auction sell much higher or lower than the same type objects at shops and shows depending on the particular attendance at the auction. Remember there is never a "set" price for any antique or collectible, and compared with many areas of collecting—this is especially true for brass because it is diversified in so many ways. Ultimately it is up to the individual to decide if the price is right. In essence, it is the individual as well as the dealer who set prices based simply on the law of supply and demand.

Useful Terms

When reading and studying about various subjects, certain technical words are often used which are petinent to the specific subject. Sometimes a word may have a different meaning than the usual definition of the word when used in connection with some specific topic. The following list of words with brief definitions may be encountered in the study of brass. These words deal with the metal itself and relate chiefly to either the manufacture or decoration of the metal. Such processes can be explained more easily in this section than in the text.

Alchemy--the Medieval form of chemistry where elements, especially various metals, were mixed together in the search for gold. The practitioner of this was known as an "alchemist."

Alloy--the combination of two or more metals to make another metal. Alloys are made for various purposes such as to increase strength and durability of a metal, change its appearance, or make the base metal less expensive.

Battery Works--an early name applied to the location where brass was made and shaped into objects.

Beaten Brass--refers either to the process of beating the molten brass into sheets or into objects in past centuries. The term may also refer to designs made on brass objects by hammering.

Brass--a metal alloy composed of copper and tin and very ductile in nature.

Brasses--this word may describe small items made of brass, but it frequently refers to monuments for graves used in England during the Middle Ages which were made of brass.

Brazier--may refer to persons who made brass or to containers, sometimes made of brass, for coal-made fires.

Brazing--the technique of joining metals together with a metal or a metal alloy.

Bronze--a metal alloy composed of copper and tin.

Cast--to form or shape an object by using a mold.

Chasing--a method of decorating brass with designs in relief by the use of tools to chisel or hammer the metal.

Core Casting--a technique of casting brass is one piece with a holow core rather than casting an object into halves which must be soldered together.

Corrode--in reference to brass, the word means to destroy the metal through contact with a chemical agent.

Distill--a process to extract one substance from another such as extracting the element of pure zinc from its ore.

Die--a metal press used for stamping shapes and designs on brass.

Doré--a French term for gilding brass or bronze. The English word for this method of decoration is "ormolu."

Dovetail--a method of construction where the metal is cut with extensions on one part which fit into similar open spaces of another part. Evidence of this type of work can be found on some brass items such as kettles and pans. The two parts would have been soldered together.

Ductile--malleable or flexible, easy to shape and able to accommodate stress during the shaping process.

Electroplating--a process of covering an object, frequently brass or copper, with a metal coating (often silver) accomplished by using electric currents.

Element--a substance found in nature which cannot be chemically decomposed such as gold, copper, silver, zinc.

Embossed--designs applied to brass by hammering the pattern from the inside of the object or applying a design to the exterior to achieve a raised effect. The French term for this type of work is "repousse"

Enamelling--a decorative technique that uses a vitreous substance in various colors which is applied by fusion to brass objects.

Engraving--a method of decorating brass by cutting or carving designs into the metal.

Flat Ware--normally means objects without a hollow center such as forks, knives and spoons as well as objects with flat surfaces such as plates and trays.

Friable--easily broken, will not tolerate stress. Bronze is friable whereas brass is not.

Fusion--the method of blending two elements together to make one, such as fusing copper and zinc to make brass.

Gilt or Gilding--the process of applying a thin coating of gold which has been mixed with mercury and then fired onto the surface of a brass or bronze object.

Hammered Brass--this term is similar to "beaten brass," indicating that the metal was either hammered into shape or that various designs were hammered into the brass.

Hollow Ware--refers to objects which have a hollow center such as pitchers, tea and coffee pots, cups, etc.

Lacquered--the application of a thin coat of varnish to the surface of a brass object to prevent the metal from tarnishing and becoming discolored.

Latten--an old word used to describe sheet brass.

Lapis Calaminearis--the natural ore containing the metal zinc.

Lost Wax--a method of casting brass in Medieval times used through the 17th century. In this process, a model of the object was made of wax and then covered with clay. After the clay mold had hardened, it was heated at a high temperature, thus melting or "losing" the wax mold. The clay mold was then filled with molten brass. When the brass had cooled, the clay mold was broken away leaving the brass object. The French term for this process was "Cire Perdue."

Metallurgy--the science of metals, or making metals from alloys, or extracting metals from their natural ores.

Mold or Mould--to shape an object. For brass, molds were made of clay, wax, and sand.

Molten--the state of a substance melted by high heat.

Ormolu-gilded pieces of brass or bronze used to decorate furniture. Sometimes the brass alloy was applied as decoration without gilding because of its own resemblance to gold in color. The French term for this method of decoration is "doré."

Patina--the surface of an object such as brass that darkens with time and exposure to the air. A patina may also be achieved through the results of polishing over time. Brass may also acquire a patina by being handled over a long period without any polishing.

Pierced--a method of decorating brass with open-work designs punched into the metal.

Plated--to cover a metal or metal alloy with another metal or alloy. Brass was often plated with silver, and other materials have been plated with brass (see Electroplating). Plating allows only a small amount of the more expensive metal to be used over a cheaper metal while giving the appearance that the item is made entirely of the more expensive metal.

Punched--a technique of decorating brass by a tool that perforates the surface or may just indent the surface without perforation.

Repoussé--designs made in relief on brass (see Embossed).

Sand Casting--a method of casting brass which largely replaced the "Lost Wax" method during the 18th century. The technique is still in use today, especially for casting bronze. A wooden mold is made and placed in a special container which has been filled with a wet sand substance. The sand is packed tightly around the mold. After the wooden model has made its impression into the sand, it is removed. The sand must become dry and harden; then the molten metal can be poured into the mold, eventually resulting in the metal object.

Sheet Brass--molten brass which, after cooling, was beaten into thin sheets by hand at first and later by water-powered hammers until the rolling machine was invented during the early 1700s.

Smelt--to melt metals or ores containing metals, or to make alloys by melting two or more metals together such as copper and zinc to form brass.

Stamping--the process of making either designs on brass or stamping out the object completely by using a metal die.

Solder--to join together with a metal alloy or metal, such as soldering the sides of a candlestick.

Solid Casting--the shaping of a brass object without a hollow core.

Spun Brass--a technique of shaping brass with the use of a die and rotating device resulting in lighter weight brass objects such as kettles and pans.

Tarnish--refers to a metal losing its shine and becoming discolored, as though oxidation.

Tin--a metal used alone, or with other metals to make alloys. For example, copper and tin are combined to make bronze. Tin was also used to line brass cooking ware and eating utensils in the past because it prevented a "tainted" taste from direct contact of brass with food; also, it kept acidic foods from corroding the brass.

Zinc--a natural element used with copper to make the alloy brass.

Lighting Implements

PLATE 1. Candleholder, 7″h, 2½″d of octagonal sided base, English, ca. mid 1800s, one of a pair.

PLATE 2. Chamberstick, 7¼"l, heart shaped, English. Chambersticks were carried to light the way to bed or up the stairs. Many were simply round in shape and made of other material as well as brass.

PLATE 3. Candleholder, 7"h, 6½"d, French, ornate styling in scroll work connecting top of holder to base. The base rests on three rounded feet, one of a pair.

PLATE 4. Candleholder, 47″h. This is a floor-style or altar type holder often used in churches. The candle rests on the spike which is called a "pricket." The "pricket" style is one of the earliest forms of candleholders. This example is French ca. early 1800s. Note the paw feet and applied cherubs at the base.

PLATE 5. Candelabrum, 12″h, 5 light, Art Nouveau style, one of a pair.

PLATE 6. Candlestick, 7¼"h, Queen Anne period, ca. 1710, the seam is faintly visible down the side, the base is octagonal with deeply notched corners.

PLATE 7. Candleholder, 10"h, base is 4½"x3½", English, early 1800s, one of a pair.

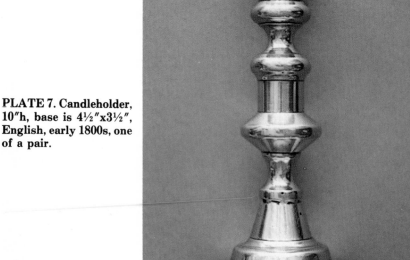

PLATE 8. Candleholder, 8″h, 3½″ base, inverted baluster stem, English, mid 1800s, one of a pair.

PLATE 9. Candleholder, 9½″h, 4″d base, English, mid 1800s.

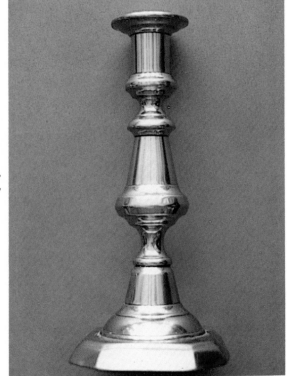

PLATE 10. The assortment of candleholders in photographs 10 and 11 are Salesmens' Samples. Pair of candleholders with hexagon sided bases, 3½"h; two light holder, 2½"h.

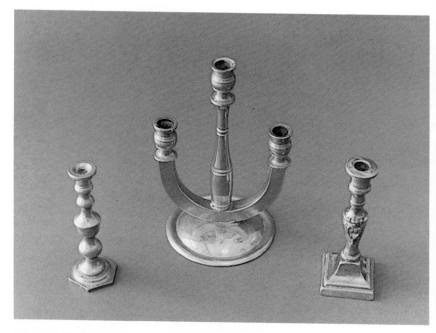

PLATE 11. Candleholders, Salesmens' Samples. The holder on the left with hexagon sided base is 3½"h; right with square base, 3½"h; three-light candelabrum, center, 4½"h.

PLATE 12. Candleholders, 8"h, deep drip pan below nozzle of holder.

PLATE 13. Candleholders, 10"h, American, 18th century. These holders are equipped with "push-ups," which were bars inserted in the stems which could be "pushed-up" to eject the candle stub after it had burned down.

PLATE 14. Candleholders in a "barley twist" design, marked "Made in England."

PLATE 15. Pair of Peg Lamps, French, enamelled glass bases, set in a pair of square based brass candleholders. The Peg Lamp is a scarce item. The glass font burned oil and was made with an extension on the base to fit into the candleholder, ca. early to mid 1800s.

PLATE 16. Photographs 16 through 19 are examples of Wall Sconces. This type of lighting has been popular for centuries. Sconces were originally made of wood and then iron, mounted on walls to hold candles and furnish light. During the 20th century wall sconces were wired for electricity and equipped with "false" candleholders to hold light bulbs which are often shaped in the form of a flame. The Wall Sconce in this photograph is very rococo in design and features a mirror in the center, English, Victorian, one of a pair, 18½"l, 13"w.

PLATE 17. Electrified Wall Sconce, 11"l, circa 1920s, one of a pair.

PLATE 18. Wall Sconce, 15″w, 19″l, heavily embossed designs featuring a bird and scenic decor, English.

PLATE 19. Wall Sconce, 26″l, 18″w, form of "Medusa" in high relief, English, ca. 1830.

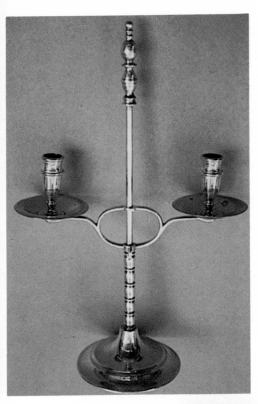

PLATE 20. Candelabras (Candelabrum indicates only one) were designed to provide more light at the table and thus vary in the number of candles they hold. They were originally made in pairs, but often only one reamins of that pair today. Because dining tables are not as large, however, one is often sufficient and can be used on many other pieces of furniture as well such as pianos, desks, and mantles for example. The one in this photograph is one of a pair, 16″h, 10″w, a simple but elegant design, English.

PLATE 21. Candelabrum, 20″h, 16″w, 3 light, English, ca. mid 1800s.

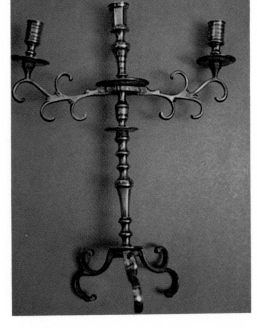

PLATE 22. Candelabrum, 4 light, marked "P" in a triangle and made by Pairpoint Manufacturing Co. (New Bedford, Mass.), early 1900s. This item was originally silver plated. Engraved floral design on center stem with "torch flame" finial.

PLATE 23. Candelabrum, 16"h, 5 light, embossed floral designs on base and drip pans, Art Deco style, ca. 1930s, originally silver plated.

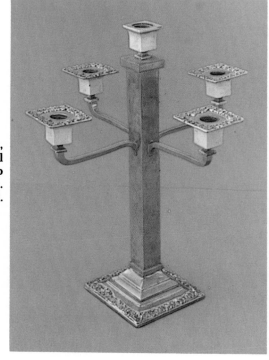

PLATE 24. Campaign Torch, 15″h (mounted on wooden stand for display). This type of torch was carried by hand in parades during the latter part of the 19th century. It has a spring device inside the stem which pushes up the candle as it burns down. The glass shade (on the side) covers the candle fitting into the socket at the top. These are sometimes mounted on walls today inside or outside the home as unique forms of lighting.

PLATE 25. Kerosene Lamp, 10″h, new satin glass shade.

PLATE 26. Aladdin Lamp Base, marked "Mantle Lamp Co., Chicago, Made in U.S.A.," 12½ "h. Many of these lamps were nickel plated and today have been stripped to the base metal of brass or copper. Price increases if lamp has original shade.

PLATE 27. Student Lamp, 20½"h, reservoir for oil on the side. The reservoir and font slide up and down to adjust the height of the light. This one has been electrified and lacquered.

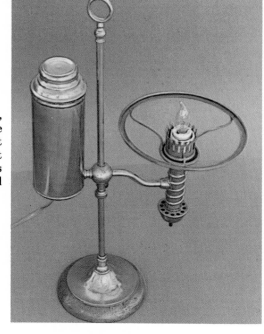

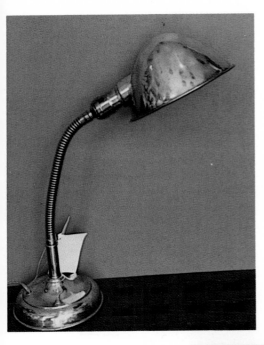

PLATE 28. Goose-Neck Lamp, electric. The neck bends to direct the light.

PLATE 29. Desk Lamp, 20"h, electric, art glass shade, neck adjusts by key. The art glass shade increases the price of this lamp.

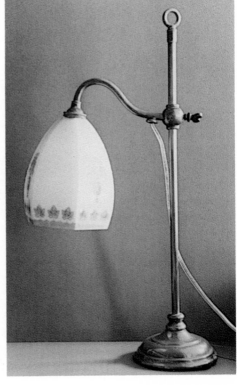

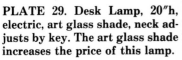

PLATE 30. Table Lamp, electric, 17"h, ornate design of Dragon body with lady's head, original silk shade, English, ca. 1920s

PLATE 31. Table Lamp with slag glass inserts in brass shade, silk fringe. Art Deco style, electric.

Fireplace Equipment

PLATE 32. Fire Dogs are shown in photographs 32 through 35. These are a type of andiron where the logs were laid across the pair. Some were simple and others quite ornate. The ones in this picture are a combination of brass and iron, 7"l, 5"h, English, ca. mid 1800s.

PLATE 33. Fire Dogs, 7"h, 8½"l, European origin.

PLATE 34. Fire Dogs, 8″h, spherical base with embossed flower and leaf designs, European, ca. mid 1800s.

PLATE 35. Fire Dog (one of a pair), 11½″h, 8½″w, European, early 1800s.

PLATE 36. Andiron (one of a pair), double spur legs, ball feet, early 1800s.

PLATE 37. Pair of Andirons, 18½″h, double spur legs, ball feet, cannonball style, American.

PLATE 38. Brass and iron Andirons, ornate pierced base with ball feet and finials, note ball finial behind shaft.

PLATE 39. Andirons, French, gilded, "Genie's Lamps with flares" distinguish this pair.

PLATE 40. Fire Screen, 18½"l, 17½"w, Georgian style, English, ca. 1870, intricate cut-out work featuring flowers, birds, leaves and a large sunburst at base.

PLATE 41. Fire Screen, masted ship in relief decor, English. This type of fire screen is typical of the ones made during the 1930s through the 1950s.

PLATE 42. Trivet designed to hang on inside of a Fender. The purpose of the trivet was to keep a kettle or pan hot over the fire, English, early 1800s.

PLATE 43. Fender, 24″l, English. Fenders were made to retain the ashes within the fireplace. This one has an attractive simple double row of pierced work for decoration.

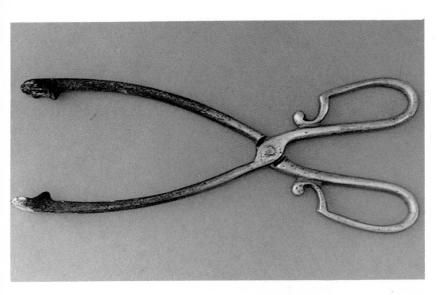

PLATE 44. Brass Coal Tongs, 10½″l.

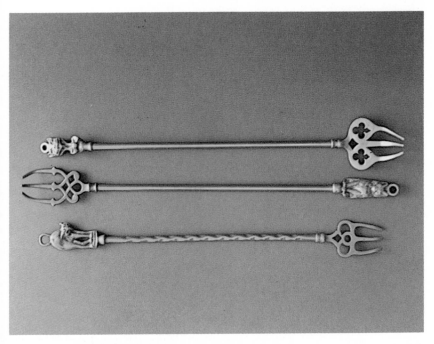

PLATE 45. Toasting Forks, 18″ to 20″ long with different pierced designs, all with figural cat handles, noted to be marshmallow toasting forks.

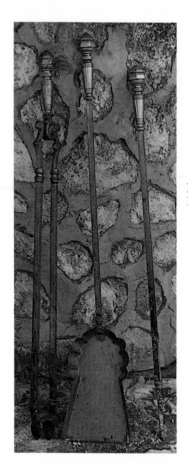

PLATE 46. Fireplace Tools: Shovel, 27″l,
Poker, 25″l, Tongs, 26″l.

PLATE 47. Photographs 47
through 52 are various containers
for coal. They may be called by
several names such as coal bucket,
coal scuttle or coal hod. The one
in this photograph is 8″h, 14″d,
English.

PLATE 48. Brass Coal Hod with copper liner, 22″h, English, early to mid 1800s, pattern of punched designs on front and sides,applied handles, footed with embossed designs. The ring visible at the back is to hold the coal shovel.

PLATE 49. Coal Bucket, embossed floral and fruit designs in high relief, English, mid to late 1800s. This piece has been worn completely through in small places around the base.

PLATE 50. Coal Bucket, 13"h, English.

PLATE 51. Coal Scuttle, 7″h, 11″d, English.

PLATE 52. Coal Scuttle, 14″h, 20″d, similar to one in Photograph 51, but much larger.

PLATE 53. Brass Box equipped with casters, English, known as a "Slipper Box." These boxes are often used to store kindling wood by the fireplace today. The examples shown here and in the next photograph were made during the first half of the 20th century, but new ones with similar designs are currently on the market.

PLATE 54. Slipper Box, 14"h, 16½"l, 11"d, mounted on casters, embossed figural tavern scene.

PLATE 55. Photographs 55 thorugh 62 are called Footmen or Trivets. Properly speaking, Trivets have 3 legs while Footmen have four legs. The trivet or footman was made for the fireplace to keep kettles or pans warm. The one in this photograph is 4½″h, 6″l with cut-out club and star shapes and heavy legs, English, ca. mid 1800s.

PLATE 56. Footman, 14½″h, 13″w. This example is much larger than the preceding one, English, early to mid 1800s.

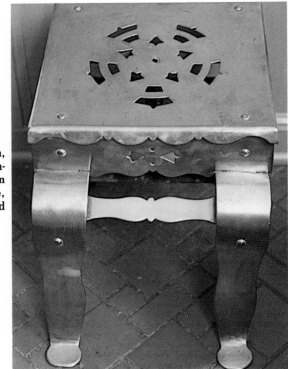

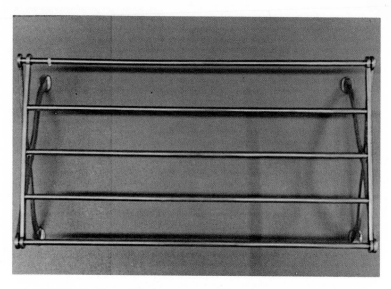

PLATE 57. Folding Fireplace Trivet, 7½″h, 24″l. This one could hold many kettles or pans, and was convenient to store when not in use.

PLATE 58. Footman, 7″h, 16″l, back legs are new, English.

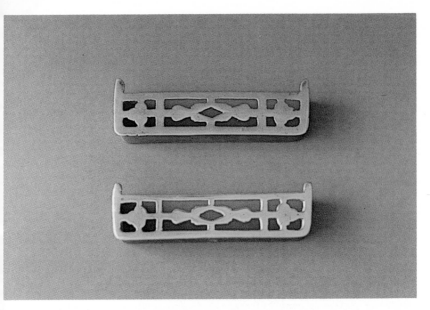

PLATE 59. Salesman's Sample of a Trivet which does not have legs, but rests on a solid strip of metal on three sides. Note the similarity of design to the one in photograph 60.

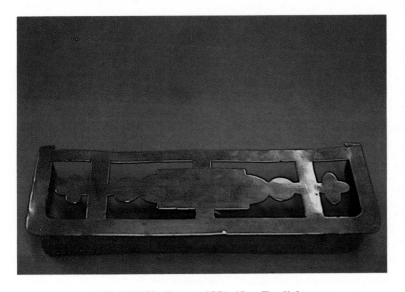

PLATE 60. Trivet, 12"l, 4"w, English.

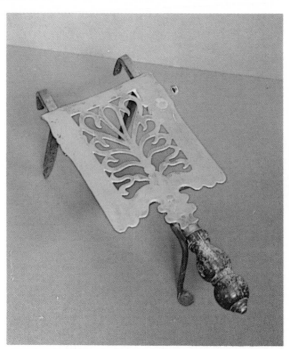

PLATE 61. Trivet, on 3 legged iron stand, wooden handle, 11½"l, 5¼"w, early 1800s, English.

PLATE 62. Footman on wrought iron stand, 15½"h, 14"w, solid sheet of brass with no piercing.

Kitchen Items

PLATE 63. Candy-Making Pan, 16″d from handle to handle. Brass allows the temperature to be the same on the overall surface of the pan which was especially important in making candy, American.

PLATE 64. Kettle or cauldron, 12″h, 19″d, iron bail, American, mid 1800s. This type was used for making apple butter.

PLATE 65. Brass Kettle for making jelly, iron bail, 7"h, 13"d, American, mid to late 1800s.

PLATE 66. Kettle, iron handle, English, tin lined, 18th century, decorated with embossed lions and shields. Notice the wear inside the kettle and the piece also has patch marks.

PLATE 67. Milk or Water Pail, 9"h, 10"d, Rim forms loops to hold bail.

PLATE 68. Kettle, 11"h, 9½"d, brass bail.

PLATE 69. Kettle for making jelly or preserves, 7″h, 12½″d, brass handle, American. This style is being copied today.

PLATE 70. Apple Butter Kettle, 10″h, 17″w, iron handle, American, note wear and mending on piece at bottom.

PLATE 71. Butter Mold of wood and brass, 6″h, 9¼″w, American, probably Minnesota or Wisconsin origin, a one pound mold.

PLATE 72. Butter Mold similar to the one in Photograph 71.

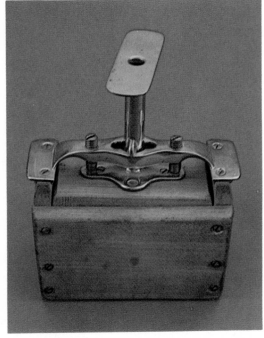

PLATE 73. Dipper, copper bowl with brass handle, another example of brass combined with another material. Brass handles are often found on copper pieces because they do not get hot as quickly as a copper handle. This dipper is 26″l, the bowl is 7½″d, mid to late 1800s.

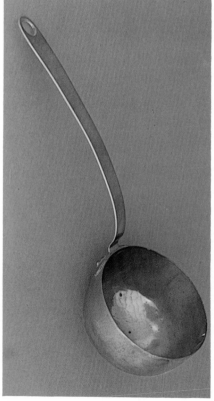

PLATE 74. Dipper, all brass, 13″l, 4½″d bowl.

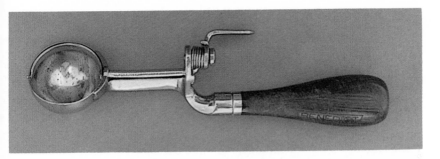

PLATE 75. Ice Cream Dipper, 10″l, brass and wood handle, bowl was nickel plated, some wear showing. Piece is marked "#6 and INDESTRUCTO."

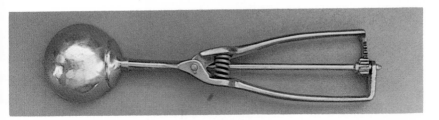

PLATE 76. Ice Cream Dipper, 10″l, brass handle with nickel plated bowl, marked "Gilchrist."

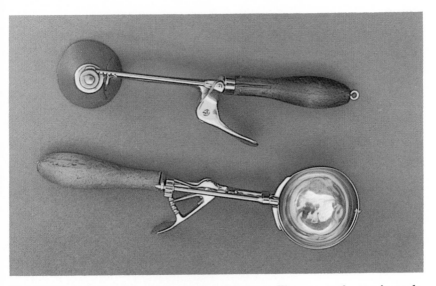

PLATE 77. Two Ice Cream Dippers by Gilchrist. The one at the top is made of brass, nickel, and wood. This particular shape formed "conettes." It is 10½″l, and marked "#33" along with the manufacturer's name. The one at the bottom is 12″l with a wooden handle. All of the original nickel plating has been stripped from the bowl.

PLATE 78. Mug, 7"h, base has been mended with copper.

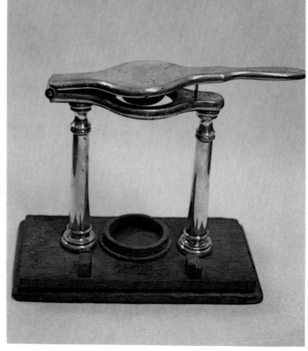

PLATE 79. Lemon Press mounted on wooden stand, English, ca. mid 1800s.

PLATE 80. Three examples of brass teaspoons, quite worn, American, 19th century.

PLATE 81. Mortar, 3¼″h, 5½″d; Pestle, 7″l. These were used in the kitchen for grinding herbs.

PLATE 82. Colander, brass pan with four circular punched designs, iron handle.

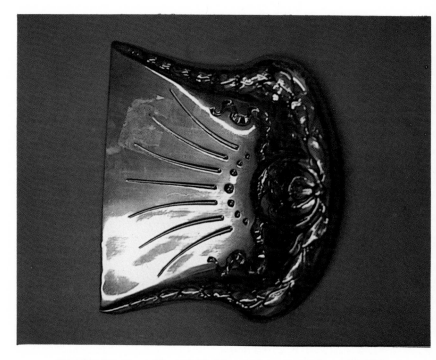

PLATE 83. Dust Pan, 8¼″l, 8″w, embossed designs, English.

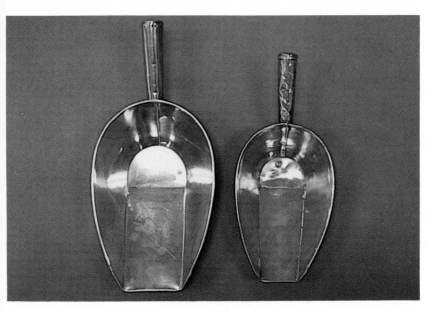

PLATE 84. Scoops for sugar or grain. These are marked "Patented Dec. 8, 1868" on handle, American.

PLATE 85. Tea Caddies, 8″h with copper labels. The one on the left is marked "Tea, Fine Sinagar," and the one on the right "Tea, Choice Assam," English.

PLATE 86. Clockwork Roasting Jack, 11″l, marked "SALTER," English. Roasting devices of this type were made during the early 1800s to cook meat in the fireplace. They were operated by inserting a key into the hole on the side which wound up a spring inside. The spring mechanism was similar to that used in clocks, and thus the name "Clockwork" became identified with the jack. "Bottle-Jack" is also a term used to describe the jack because the spring inside was shaped like a bottle.

PLATE 87. Rub Boards were essential in the home around the turn of the century. The manufacturer's name is still often visible at the top of many, although it has been worn away on this example. Rub boards of this type were selling for 33 cents in 1908.

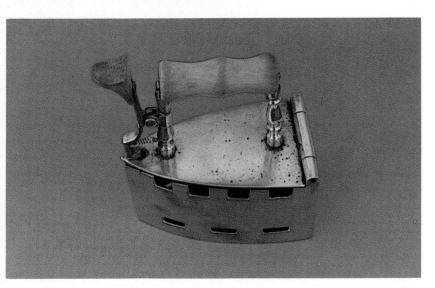

PLATE 88. Charcoal Iron, 8″h, hinged top, wooden handle has been replaced. Variations of this type of iron were used in 19th and early 20th century American households. Hot coals were placed inside the iron.

PLATE 89. Brass Kettle with copper spout, 10″h, 12″d, English.

PLATE 90. Kettle, 12″h, goose-neck spout, rounded brass handle.

PLATE 91. Kettle, 8″h, wooden handle and finial. This kettle is early 20th century while the preceding ones are from the 19th century.

PLATE 92. Watering Can, 8″h, 11″w, hinged lid, European.

PLATE 93. Tea Pot with stand and burner, marked "SAS" in a circle, early 1800s.

PLATE 94. Tea Pot, rounded body, wood and brass handle, with 3-legged stand or trivet, English, early 1800s.

PLATE 95. Tea Pot, footed, amber glass handle, English, ca. mid 1800s.

PLATE 96. Coffee Service, Art Deco style, ca. 1930s, marked "Doryln Silversmith."

PLATE 97. Coffee Pot, Dragon Eye spout, names are engraved on each side, "CARAPLE" and "AMALIE," circa early 20th century.

PLATE 98. Samovar, 25″h overall, marked "Tehran."

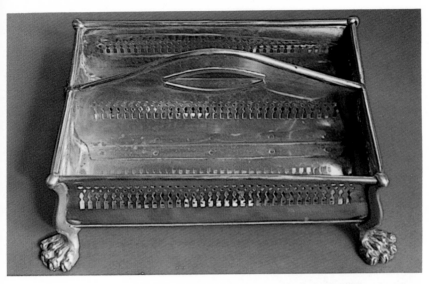

PLATE 99. Knife Box or Tray, English, Georgian period, early 1800s, claw feet and pierced designs.

PLATE 100. Wine Cooler on pedestal base, 10″h, 8″d, applied handles of lion heads with ring though noses, mid 20th century.

PLATE 101. Crumber for cleaning the table. These usually had a matching tray.

PLATE 102. Crumber Tray and Holder.

PLATE 103. Creamer and Sugar, brass with pewter handles.

PLATE 104. Tray, 12″d, American, marked "S. & S. H.," originally silverplated, mid 20th century.

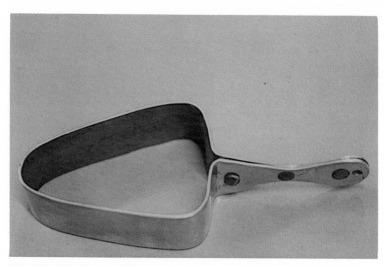

PLATE 105. Trivet, 10"l, 3¼" wide, hollow center. This style of trivet could be used to hold an iron or used at the table for hot dishes.

PLATE 106. Tray, 11"x16", stippled design, mid 20th century.

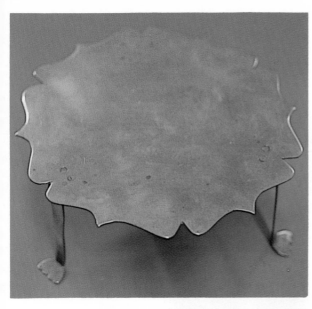

PLATE 107. Trivet, 3½"h, 8"d, fancy scalloped border, smooth top, early 1800s.

PLATE 108. Trivet, 7¼"l, fox and tree design.

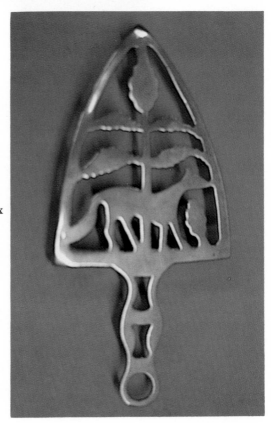

PLATE 109. Trivet, 9¼″h, 9″sq., elaborate pierced work, English, ca. mid 1800s.

PLATE 110. Trivet on pedestal stand, 9½″h, 7″d, design is more simple on this trivet than the preceding one, some wear on top.

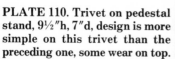

Tools and Instruments

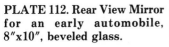

PLATE 111. Photographs 111 through 114 are of objects connected with various vehicles. Coach lantern, 18"l, mid 1800s.

PLATE 112. Rear View Mirror for an early automobile, 8"x10", beveled glass.

PLATE 113. Trouble Light, 8″l, convenient for providing light under hoods or vehicles.

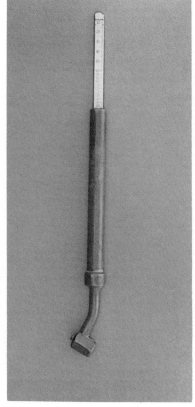

PLATE 114. Truck Tire Air Gauge, marked "A. Schrader's Son, Service Tire Gauge Division of Scovill Manufacturing Co., Brooklyn, N.Y."

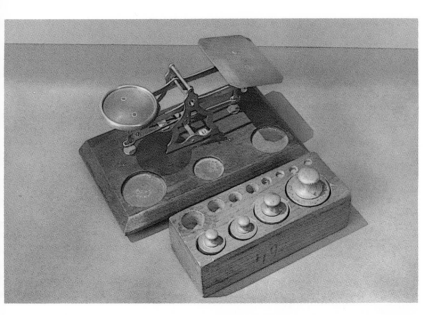

PLATE 115. Photographs 115 through 125 are examples of scales and weighing devices. Scales range from small ones used by post offices, pharmacies, and jewelers to larger types used by merchants for weighing candy, grain, meat and produce. This set is 8″l, 4″w, European.

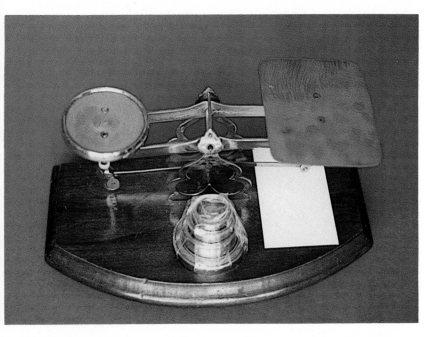

PLATE 116. Wooden and Brass Scales, 9″l, brass weights, European.

PLATE 117. Merchant's Scale, marked "J. Hart, Maker, Birmingham," English, mid 1800s.

PLATE 118. Round Scale with brass plate to calibrate pounds, marked "Chatillon."

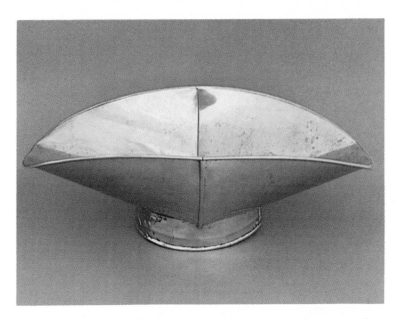

PLATE 119. Pan or "Scoops" used on spring or weight-type scales, American. This one has a pedestal base, 20"l.

PLATE 120. Balancing Scales, 11"l, French, mounted on walnut base.

PLATE 121. Scale, 13″l, English, marked "W. & T. Avery, Ltd., Birmingham."

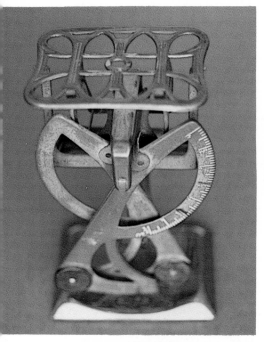

PLATE 122. Postage Scale, markings illegible.

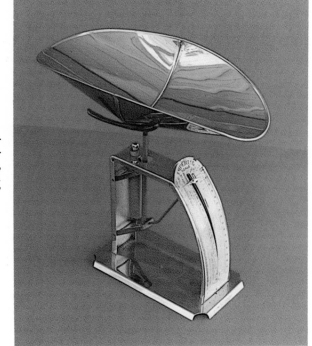

PLATE 123. Merchant's Scale, marked "Imperial Scale, Gilfillan Scale & HDW Co., Chicago," 9½"h.

PLATE 124. Measure for Grain, 8½″h, 4½″d, marked "HOWE."

PLATE 125. Scale, 6½″h, measure, 9″l. The measure fits into the cylinder of the scale, rope handle, marked "Swedish Made."

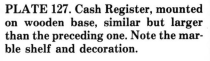

PLATE 126. Cash Register, 12″w. This one was made by the "National Company," an American firm founded around the mid 1880s. Most examples of cash registers found on the market today were made by National. Various sizes, often ornately decorated are seen.

PLATE 127. Cash Register, mounted on wooden base, similar but larger than the preceding one. Note the marble shelf and decoration.

PLATE 128. Microscope, 17″h, inscribed "BAKER, 244 High Holborn, London," ca. 1840.

PLATE 129. Miniature Microscope, 3″h.

PLATE 130.
Telescope on tripod,
English, early 1800s.
This is a day or
night type, marked
"Spencer Brown-
ing," a well-known
optician of the
period.

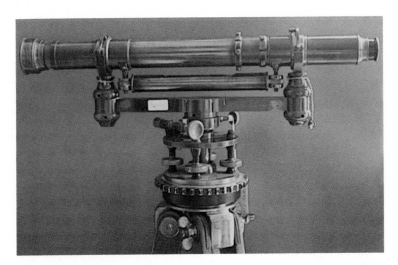

PLATE 131. American Survey Instrument, 17"l, inscribed "HERM
PFISTER, Cincinnati."

PLATE 132. Tool Holder made of brass.

PLATE 133. Cobbler's Tool, red brass Shoe Mold, 11″x9″.

PLATE 134. Chisel, red brass, 7½"l, marked "Beryl Co., S. 108," and "BE Co."

PLATE 135. part of a shoe-shine stand, 15"h.

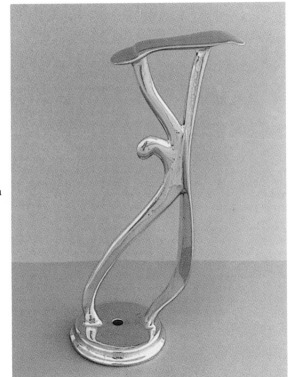

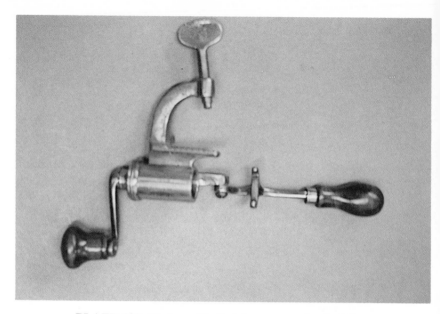

PLATE 136. Shotgun Shell Crimper, wooden handles.

PLATE 137. Fishing Reels, ca. mid to late 1800s.

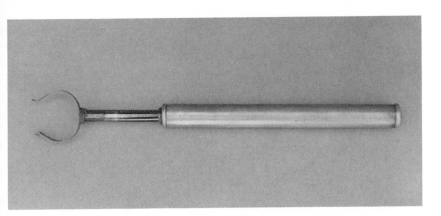

PLATE 138. Test Tube Holder, 8″l.

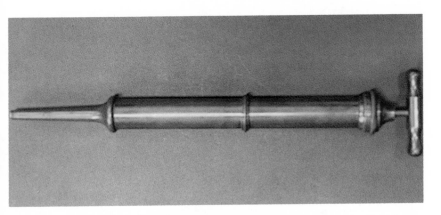

PLATE 139. Pump, 16½″l.

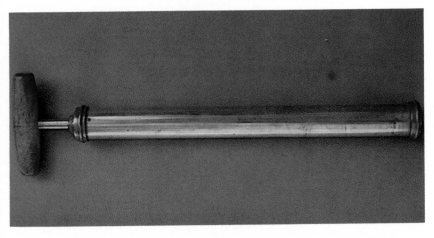

PLATE 140. Pump style Mister, marked "Whitney, Boston, Mass."

PLATE 141. Blow Torch, 7″h.

PLATE 142. Blow Torch, 11″h, marked "The Turner Brass Works, Sycamore, Ill." Trade Mark of a figure of a lady on a gym bar.

PLATE 143. Lantern, 10″h, pierced at top and bottom to allow air to enter, wooden handle.

PLATE 144. French Carbide Lantern with copper cover, 11½″h, marked "C. Quvrard & Cie."

PLATE 145. Bracket Candleholder used on railroad cars and mounted to the wall. Originally this holder would have had a glass chimney. It is 8"h. Such holders are often used as sconces today.

PLATE 146. Lantern, marked "Dietz, New York, U.S.A., 1900," 13"h.

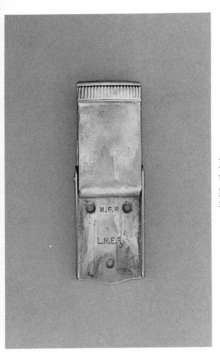

PLATE 147. A railroad collectible in the form of an Ash Tray for a passenger train, 5″l, attaches to seat, marked "N.F.R." and "L.N.E.R."

PLATE 148. Keys to Railroad locks, from left to right: marked: "NEV CRR"; "CPRR of CAL"; "ETH & CRR, W. Bomannan, Brooklyn, New York."

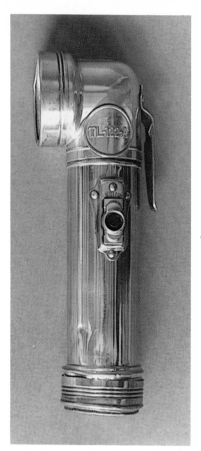

PLATE 149. Flash Light, nickel and brass, 7″l, marked "TL-122-A," ca. 1930.

PLATE 150. Coal Miner's Head Lamp, 4½″h, marked "Made in USA, Pat. Pending," by "JUSTRITE," and "United States of America" on base.

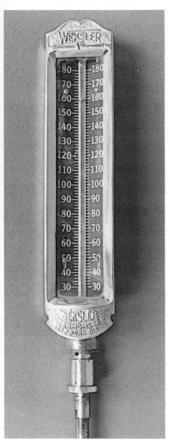

PLATE 151. Thermometer, 13″l, marked "Weksler, New York City."

PLATE 152. Steam Whistle, 9¾″l, marked "Kinsley Mfg. Co., Bridgeport Co., USA."

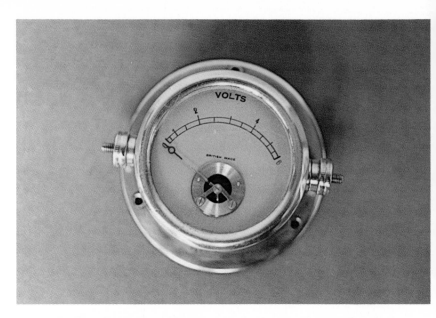

PLATE 153. Voltage meter, marked "British Made," 1¾"x3½".

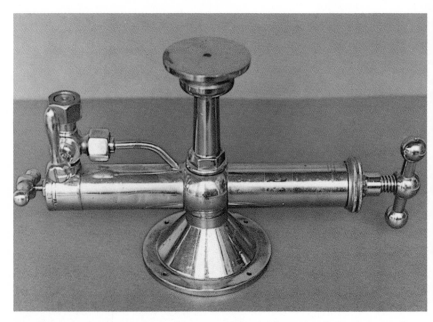

PLATE 154. Pressure Gauge Tester, marked "Schaffer & Budenberg Corp, Brooklyn, N.Y." 13½"l, 8¼"h.

PLATE 155. Brass Shell, 6½″l, 2″d, ca. 1915.

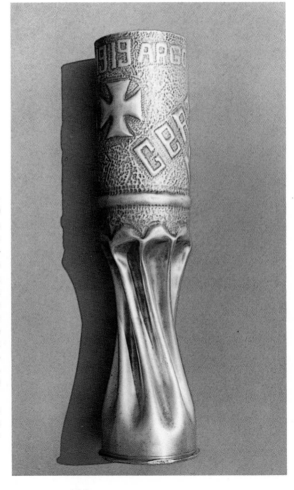

PLATE 156. Photographs 156 through 160 show examples of brass tools and instruments which have been made into other articles. The vase in this photograph is an example of "Trench Art." The vase is made from a shell similar, but larger than the one in the preceding photograph. It is 13½″h, made by a German jeweller. The wording on the vase is "1919 Argonne, Germany." Work of this sort is especially of interest to collectors of military artifacts.

PLATE 157. Brass Shell, 2 feet long, 105 milimeter, made into a table lamp.

PLATE 158. Fire Extinguisher made into a floor lamp.

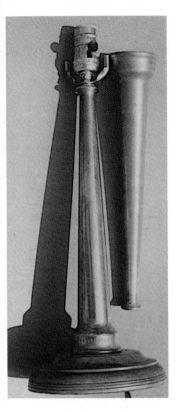

PLATE 159. Hose Nozzle from a fire truck, shown alone and made into a lamp. The nozzle is marked "Elkhart Brass Mfg. Co., Indiana."

PLATE 160. Table made from a brass Port Hole from a World War II Merchant Ship.

PLATE 161. Photographs 161 through 184 illustrate "Nautical" antiques and collectibles. Many of these date from the World War II Period. A Buoy Safety Light is shown here.

PLATE 162. Ship's Bell, 9″h.

PLATE 163. Yacht Wheel, brass and mahogany, 28″d.

PLATE 164. Ship Reflector, English, marked "Kelvin & Hughes, Ltd., NKII REFLECTOR BINNACLE," 16″h, ca. 1945.

PLATE 165. Propeller.

PLATE 166. Fog Horn, 24″h, marked "Original Makrofon."

PLATE 167. Brass Signal Cannon (in new carrier), 16″l, marked "1880, Big Boom."

PLATE 168. Dry Compass, 11½″x6½″, marked "Kelvin & James White, Ltd., 13 Cambridge St., Glasgow."

PLATE 169. Life Boat Compass, 12″x8½″, marked "BERGEN-NAUTIK."

PLATE 170. Sextant, Scottish with original label in top of case, World War II period.

PLATE 171. Sextant, U.S. Navy, World War II.

PLATE 172. Clinometer, brass in mahogany case, 8″x6″, marked "45 Degrees, Admiralty Pattern, E.R. Watts & Son, London, No. 24288," World War II period.

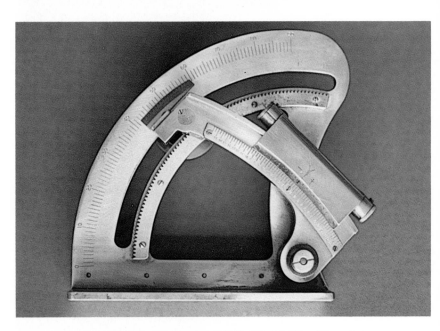

PLATE 173. Clinometer, 6¼″l, World War II period.

PLATE 174. Magnifier, 4″h, 9″d.

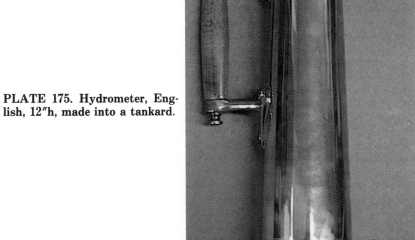

PLATE 175. Hydrometer, English, 12″h, made into a tankard.

PLATE 176. Yacht Tie-Down, 5″l.

PLATE 177. Yacht Tie-Down, 6½″l.

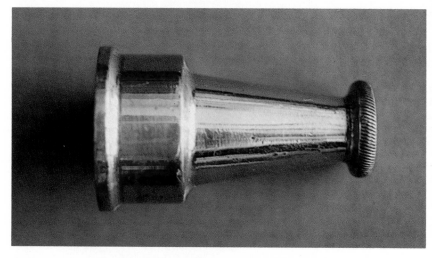

PLATE 178. Nozzle (fits on end of hoses), 3″h.

PLATE 179. Triplex Lens in brass, for bow and starboard sides of a ship, 9″l, marked "TRIPLEX," patented December 24, 1910.

PLATE 180. Brass Lantern with green glass indicating starboard or left light, marked "PERKO, Brooklyn." This lantern was made prior to World War II as the Perko Company moved to Florida during the war years.

PLATE 181. Mast Head Lamp, 10"h, made by "A. Ward Hendrickson & Co., Inc., Brooklyn, New York." Note that the glass is supposed to be white but has turned purple from age and lead content.

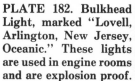

PLATE 182. Bulkhead Light, marked "Lovell, Arlington, New Jersey, Oceanic." These lights are used in engine rooms and are explosion proof.

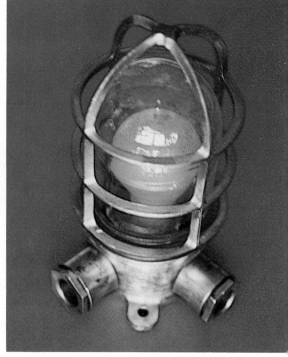

PLATE 183. Anchor Lantern, 12″h.

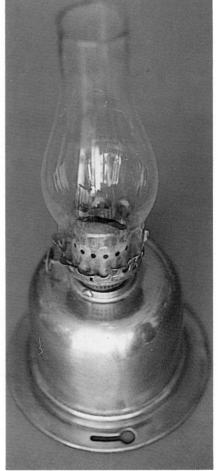

PLATE 184. Ship's Kerosene Lamp, 4½″h (base), 12″h overall, note base in pierced for attachment to a flat surface.

Hardware Fixtures

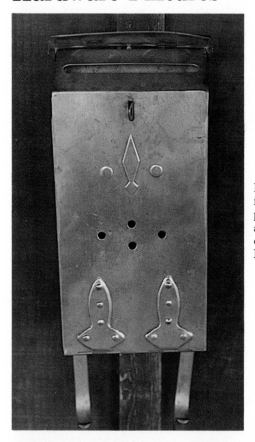

PLATE 185. The Hardware items featured in this section were made prior to the mid 20th century, although some of the styles are currently being copied today. Mail Box, 13"l with slot and key hole.

PLATE 186. Mail Box, envelope style with lion's head opener, ca. mid 20th century.

PLATE 187. Name Plaques for doors, 6″l.

PLATE 188. Commercial Door Handle, 14½″l.

PLATE 189. Door Knocker, 6″l, an ever popular style of lion's head with ring through nose.

PLATE 190. Door Knockers, 6½″l, in Art Deco style.

PLATE 191. Door Handles: Left, wood and brass, 10½″l; Right, ornately patterned handle, 11″l.

PLATE 192. A selection of Door Knobs.

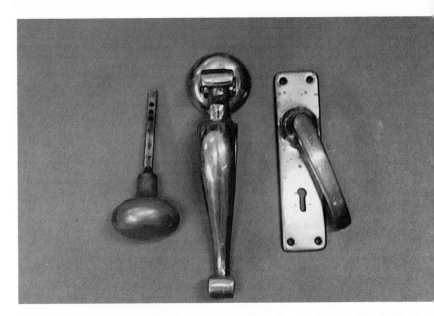

PLATE 193. Three early 20th century Door Handles: Left, Door Knob; Center, Door Handle and Knocker; Right, Door Handle with key hole opening.

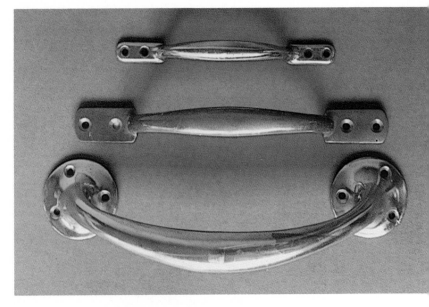

PLATE 194. Pulls for cabinets or chests.

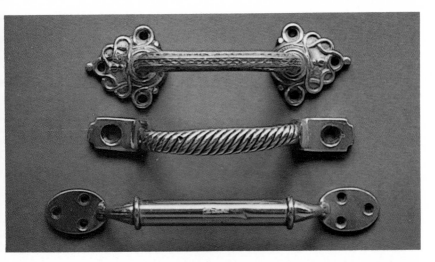

PLATE 195. Pulls for cabinets or chests. Price depends on size and ornateness of design.

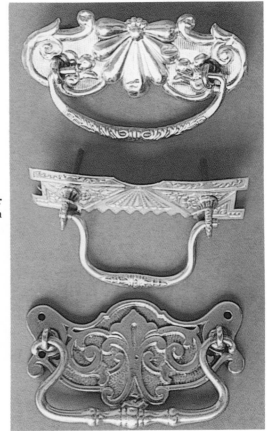

PLATE 196. Pulls for cabinets or chests with backplates.

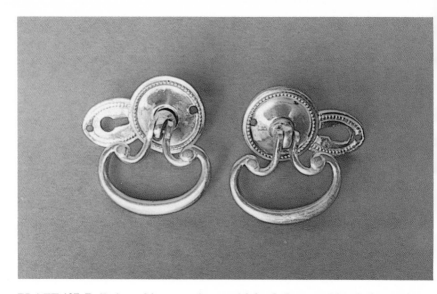

PLATE 197. Pulls for cabinets or chests with backplates and key hole openings.

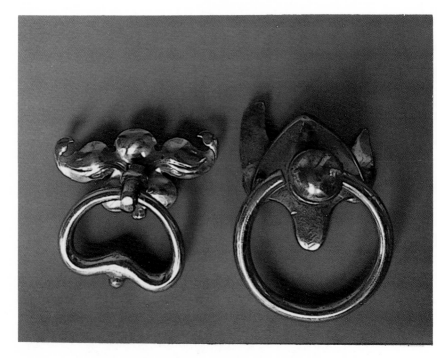

PLATE 198. Red brass Pulls for cabinets or chests.

PLATE 199. Ormolu (Furniture mounts), French. Figural design on left, 11″l; Floral and Fruit garland design on right, 10″l.

PLATE 200. Ormolu, French, figural designs: Top, 3½″l; Bottom, 7½″l.

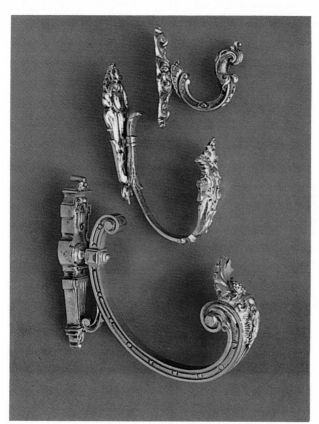

PLATE 201. Curtain Tie-Backs, showing one each of a pair.

PLATE 202. Curtain Tie-Backs: Left, knob style; Center, Art Deco design; Right, beaded decor on base.

PLATE 203. Curtain Tie-Backs in three different designs.

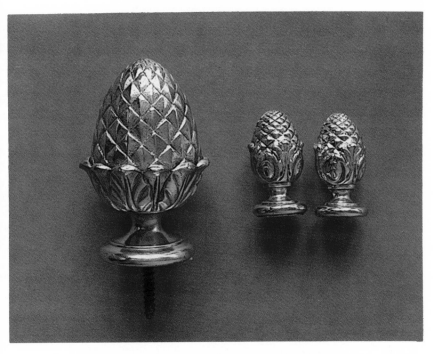

PLATE 204. Curtain Tie-Backs, pineapple design.

PLATE 205. Brass Shower Ring and Head.

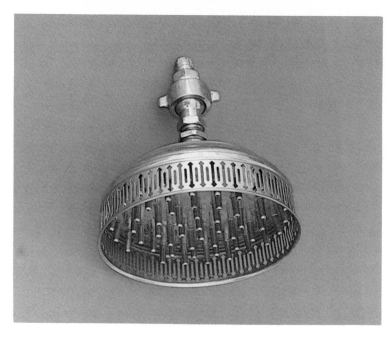

PLATE 206. Shower Head, 9½″d.

PLATE 207. Bathroom Cup Holder and Soap Dish, 8″l overall.

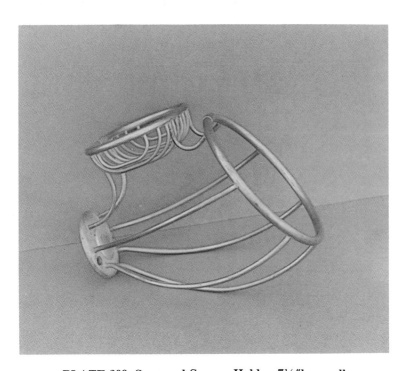

PLATE 208. Soap and Sponge Holder, 7½″l overall.

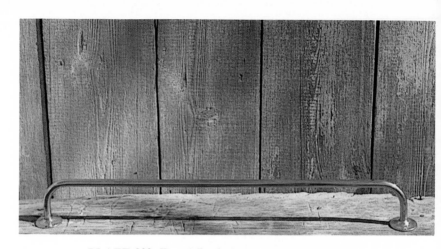

PLATE 209. Towel Rack from railroad car, 33½"l.

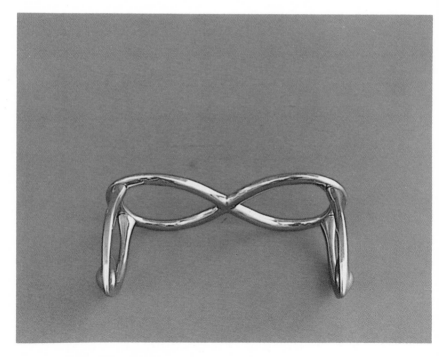

PLATE 210. Toilet Tissue Hanger, 6½"l, marked "SAN-O-LA."

Decorative and Personal Objects

PLATE 211. All types of brass bells are very popular collectibles. Although they served a useful purpose in the past, most are rarely "rung" today! The hand bell in this photograph is English and has been lacquered.

PLATE 212. Servant's Bell, 13"l, English, early to mid 1800s. This type of bell was rung by a cord connected by a wire to various rooms in the house. A set of these was used "below stairs" and would ring when the cord was pulled "upstairs." Each bell had a different sound.

PLATE 213. School Bell, 12"h overall, wooden handle. Many reproductions of school bells are made today.

PLATE 214. School Bell, 6"h.

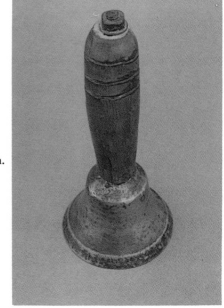

PLATE 215. Call Bell for table, store, or hotel desks.

PLATE 216. Door Bell, 7″d with cast iron back, marked with patent dates of 1872, 1873, 1874.

PLATE 217. Bed Warmers were extremely useful in the past. They were filled with hot coals and passed back and forth under the bed covers (hence the long handle). This one is 45"l, pan is 12"d with pierced design, English.

PLATE 218. Bed Warmer with copper bottom and brass top, pierced design, New England origin, mid to late 1800s, 41"l, 13"d.

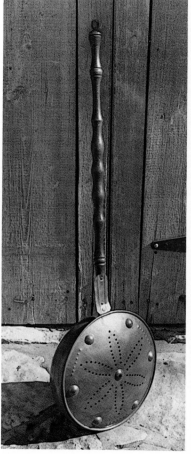

PLATE 219. Bowl and Pitcher set, Russian origin, mid to late 1800s. Pitcher sits on pierced insert.

PLATE 220. Bidet, (French, of course!), on wrought iron stand, marked "MARQUE J.B., DEPOSE," 15½"h, 21"l, 19th century.

PLATE 221. Bird Cages were very popular during Victorian times. This one is brass on a copper pedestal base in Art Nouveau style. The Cage is 41"h and the Pedestal is 39"h.

PLATE 222. Bird Cage, 39"h, 18½"d, ca. mid 20th century.

142

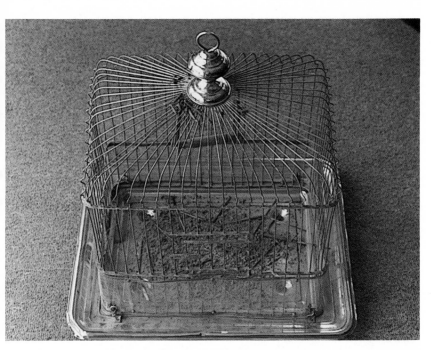

PLATE 223. Bird Cage, square shape, 16″h, with stand (not shown), ca. mid 20th century.

PLATE 224. Bird Cage, 16″h, with stand (not shown), ca. mid 20th century.

PLATE 225. Boxes made entirely of brass or decorated with brass have been made in a variety of styles and for a variety of purposes through the years. Several types are shown in photographs 225 to 232. This one is oval shaped, French, ca. 1800, 8½"l, 3"w.

PLATE 226. Bride's Box, 13" sq., 9"h, from India, ca. 1850.

144

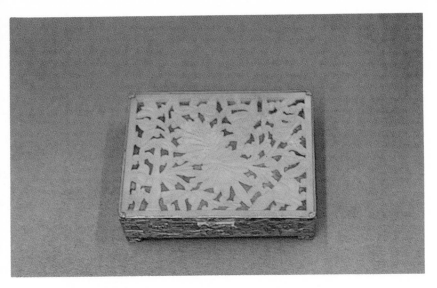

PLATE 227. Brass and Jade Box, marked "China," 6″l.

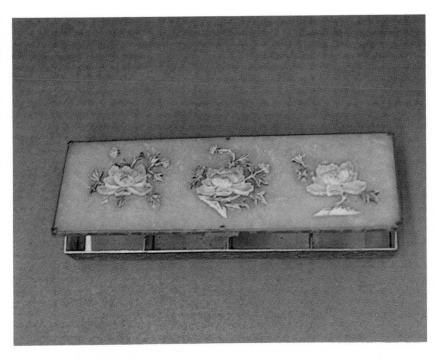

PLATE 228. Brass, Jade and Ivory Box, unmarked, 11″l.

145

PLATE 229. Brass Box, 3″h, 10½″w, wooden bottom, velvet lined, engraved figural scene on top.

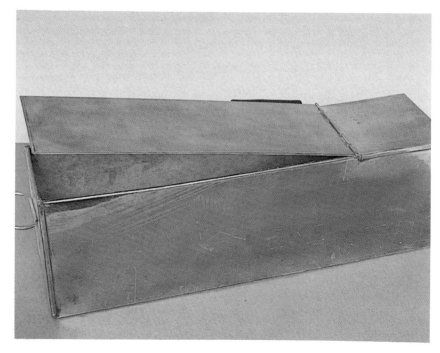

PLATE 230. Safety Deposit Box, 18½″l, 4½″w.

PLATE 231. Candy Box, made for "Schrafft's", transfer colonial design on top, 8"l, 5"w.

PLATE 232. Brass and enamel Box, 6½"l, 2½"w, Oriental.

PLATE 233. Clock made of brass, 11″h, 8″w, marked "Seth Thomas," a popular Connecticut clock company.

PLATE 234. Clock, 10″h, 5½″w, porcelain face, marked "Made in France."

PLATE 235. A number of Desk related items are shown in Photographs 235 to 246. Pair of brass Book Ends made in a Pagoda style, 6"h, 4"w, marked "China."

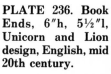

PLATE 236. Book Ends, 6"h, 5½"l, Unicorn and Lion design, English, mid 20th century.

PLATE 237. Brass Desk Accessory with Letter Holder, 7"h, 12"l, 10"d, European.

PLATE 238. Brass Crest, 12"x11", figural design of Lions and Unicorn, French inscribed *"Honi Soit Qui Mal Y Pense,"* an old French saying "Evil be to him who evil thinks."

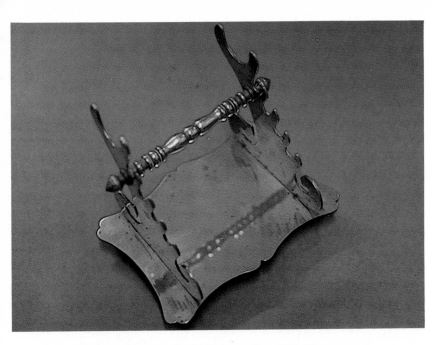

PLATE 239. Pen Staff Holder, 6″ x 6¼″, European.

PLATE 240. Brass and copper Desk Accessory with pen tip drawer, inkwell and pen holders, footed, pierced at the top for hanging, English, 8″h, 8″l.

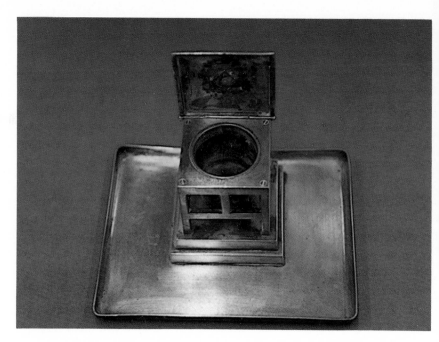

PLATE 241. Inkwell with tray, 3½″h, tray 6″l, ca. mid 20th century.

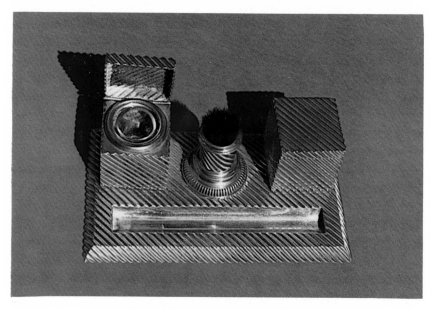

PLATE 242. Inkwell Set, 8″l, 4½″w, English.

PLATE 243. Inkwell, 5″h, signed "Stingl."

PLATE 244. Bank, English.

PLATE 245. Water Carafe, 8"h, marked "Maxwell Phillips, New York," ca. mid 20th century.

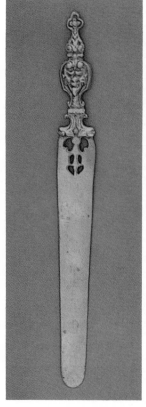

PLATE 246. Letter Opener, 10¼"l, pierced work on blade, gargoyle design on handle.

154

PLATE 247. Horse Brasses are interesting collector items which are found in quite a variety of designs as illustrated in this and the following two pictures.

PLATE 248. Horse Brasses mounted for display.

PLATE 249. Horse Brasses (for the ear).

155

PLATE 250. Censer or Incense Burner with chain, a type used in churches, intricate open-work construction, Oriental.

PLATE 251. Incense Burner, elephant motif in relief, bamboo style handles, Oriental.

PLATE 252. Jardiniere, thick ball-shaped feet, lion's head handles, 6½"h, 9½"d, English.

PLATE 253. Brass Bucket, 11"h, 10½"w, footed, applied animal designs, English, ca. early 1800s.

PLATE 254. Planter or Jardiniere, pedestal base, lion head handles, English ca. mid to late 18th century.

PLATE 255. Jardiniere, reserves decorated with engraved bird and tree designs.

PLATE 256. Jardiniere, 12″h, 17″d, lion head handles, English, early 18th century.

PLATE 257. Jardiniere, 4″h, 4″d, English, ca. mid 20th century.

PLATE 258. Jardiniere, miniature size, 2½″h, 3″d, marked "Chase." (Chase was a brass and copper company located in Waterbury, Connecticut), ca. mid 20th century.

PLATE 259. Jardiniere, 6"h, 8"d (base replaced), oval medallions with embossed scroll designs.

PLATE 260. Brass and Copper Wall Vase, 6½"l, English.

PLATE 261. Brass Wall Ornament. This is often referred to as a Comb Holder, but Schiffer (p. 22, 1978), notes that is not really the function of this piece and in fact, its use has not been actually determined. It dates from the early 19th century.

PLATE 262. Vase, 8½″h, conical shape, very heavy, Oriental, ca. mid 20th century.

PLATE 263. Jardiniere, footed, embossed grapes, English, 8"h, 7¼"d, ca. mid 20th century.

PLATE 264. Wall Pocket Vase made in form of a bed warmer, embossed fruit designs, marked "Made in England," ca. mid 20th century.

PLATE 265. Vases, leaf and branch decor in relief, 10½″h, French, Art Nouveau period.

PLATE 266. Carriage Vases, 10″l.

PLATE 267. Brass Vase Holder (would
have glass insert), 13½″h, intricate cut-out
work, ca. mid 20th century.

165

PLATE 268. Pair of Vases, 13″h, figural and floral designs in relief, twisted serpent form handles, ca. mid 20th century.

PLATE 269. Pair of Vases, 9½″h, embossed roses around center, originally silver plated, ca. mid 20th century.

PLATE 270. Shaving Mirror in brass case, 21"h, 21"w, ca. mid to late 1800s.

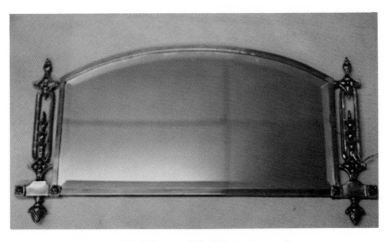

PLATE 271. Mirror, 15"x30", in brass frame.

PLATE 272. Brass Frame, 6"x8", ca. mid 20th century.

PLATE 273. Hand Mirrors with brass frames and handles. The round mirror on the left and the oval one on the right have the same figural nude design on the handle in an Art Nouveau style.

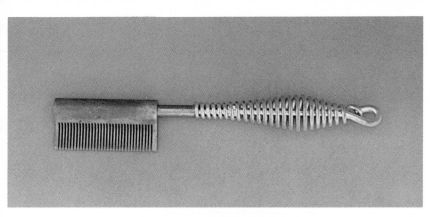

PLATE 274. Hair Straightening Comb, 10½"l.

PLATE 275. Perfume vial incased in elaborately designed brass holder.

PLATE 276. Group of Thimbles.

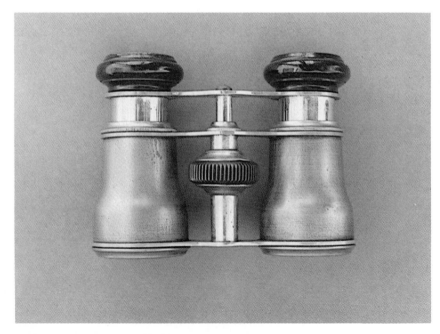

PLATE 277. Opera Glasses, 3″l, 2½″w, French, marked "Lepine, Paris."

PLATE 278. Smoking and Tobacco accessories are shown in Photographs 278 through 289. Spittoon Cover, 5"h, 7"w.

PLATE 279. Spittoon, hammered border, 5"h, 7"w.

PLATE 280. Spittoon, 3¼″h, 7¼″d, made in two pieces, marked "Farris Mfg. Co., Decatur, Illinois."

PLATE 281. Tobacco Jar, 8″h, ca. mid 20th century.

PLATE 282. Match Holder, 4½″h, ca. mid 20th century.

PLATE 283. Ash Tray, ca. mid 20th century.

PLATE 284. Ash Trays with tortoise shell trim, 3½″d. The trim increases the price of this pair.

PLATE 285. Advertising Ash Tray, "Smoke Fresh CRAVEN A," printed on leather, 4″h, 3½″d, ca. mid 20th century.

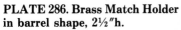

PLATE 286. Brass Match Holder in barrel shape, 2½″h.

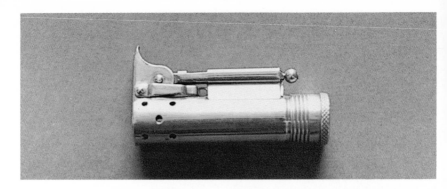

PLATE 287. Lighter, 2½″h, marked "Made in Austria."

PLATE 288. Match Holder with attached tray, 4″h.

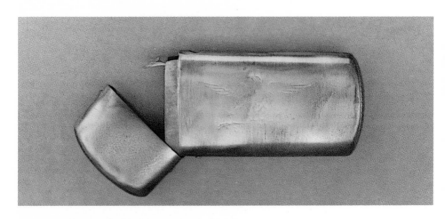

PLATE 289. Match Safe, Eagle design engraved on front.

PLATE 290. Reading Stand, adjustable, marked "S. & S. Sheldon, Birmingham," English.

PLATE 291. Pedestal or Stand, 39″h, 14″d, French, early to mid 1800s.

PLATE 292. Stool, 3 legs, 14″h, 12″d, English, chased designs on seat.

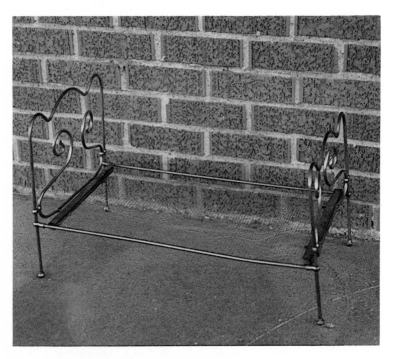

PLATE 293. Toy related item, Doll Bed, 23½″l, 13″h.

PLATE 294. Telephone, 13"h, ca. mid 20th century.

PLATE 295. Telephone, similar style as the one preceding, except this one has a brass handle.

PLATE 296. Tray, fruit shaped, 9″l, marked "Hong Kong," ca. mid 20th century.

PLATE 297. Tray in horseshoe shape, 10½″l, 7″w, footed, ca. mid 20th century.

PLATE 298. Wall Plaques, 6½"d, English, ca. mid 20th century. "Anne Hathaways Cottage" and "Robert Burns" are embossed on these souvenir plaques.

PLATE 299. White Brass Portrait Plaque, 9¼"d.

PLATE 300. Tray, 10″d, embossed scroll and floral border design.

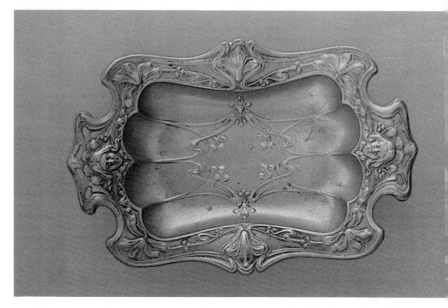

PLATE 301. Tray, 14″l, 9″d, Art Nouveau design in relief, pierced handles.

PLATE 302. Bowl, 10″d, engraved floral and leaf designs, marked "China."

PLATE 303. Compote, 7″h, 9½″d with figural copper insert of allegorical figures, French, 19th century.

PLATE 304. Umbrella Stand, embossed design of woman and two children, Dutch.

PLATE 305. Umbrella or Cane Stand, 19½"h, 25"w, French.

PLATE 306. Umbrella Stand with lion head handles, red brass, ca. mid 20th century.

Object Index to Photograph Numbers

Anchor Light--183
Andirons--36-39
Apple Butter Kettle--64,70
Ash Trays--147; 283-285;288
Bank--244
Bed Warmers--217,218
Bells--162,211-216
Bidet--220
Binnacle--164
Bird Cages--221-224
Blow Torch--141, 142
Book Ends--235,236
Bottle-Jack (see Clockwork Roasting Jack)
Bowl--302
Bowl and Pitcher Set--219
Boxes--225-232
Bracket Candle Holder--145
Bride's Box--226
Bulkhead Light--182
Butter Molds--71,72
Campaign Torch--24
Candelabra--5,20-23
Candleholders--1-16; 18-24; 145
Candy-Making Pan--63
Cane Stand (see Umbrella Stand)
Canisters--85
Carriage Vases--266
Cash Registers--126,127
Cauldron--(see Kettles)
Censer (see Incense Burner)
Chamberstick--2
Chisel--134
Cigarette Lighter--287
Clinometer--172,173
Clocks--233,234
Clockwork Roasting Jack--86
Coach Light--111
Coal Bucket--49,50
Coal Hod--48
Coal Miner's Light--150
Coal Scuttles--47,51,52
Cobbler's Tool--133
Coffee Pots--96,97
Colander--82
Comb--274
Comb Holder (see Wall Pocket Vase)
Compass--168,169
Compote--303
Creamer & Sugar--96, 103
Crest--238

Crumber--101
Crumber Tray--102
Curtain Tie-Backs--201-204
Cuspidors--278-280
Dippers--73,74
Doll Bed--293
Door Handles--188,191,193
Door Knobs--192,193
Door Knockers--189,190
Door Plates--187
Drawer Pulls, 194-198
Dust Pan--83
Fender--43
Fire Dogs--32-35
Fire Screens--40,41
Fireplace Tools--44,46
Fireplace Trivets--42,60,61 (see Footmen)
Fishing Reels--137
Flashlight--149
Fog Horn--166
Footmen--55-58; 62
Furniture Mounts (see Ormolu)
Grain Measure--124
Horse Brasses--247-249
Hose Nozzles--159,178
Hydrometer--175
Ice Cream Dippers--75-77
Incense Burners--250,251
Ink Wells--240-243
Iron--88
Jardinieres--252-259; 263
Jelly-Making Kettles--65,69
Kettles--89-91
Keys--148
Knife Box--99
Lamps--25-31; 157-159; 184
Lanterns--143,144,146,179,180,183
Lemon Press--79
Letter Opener--246
Letter Rack--237
Magnifier--174
Mailboxes--185,186
Mast Head Light--181
Match Holder--282,286,288
Match Safe--289
Microscope--128,129
Milk or Water Pails--67,68
Miniature Microscope--129
Mirrors--270-273
Mortar and Pestle--81

Mug--78
Nautical Items--160-184
Opera Glasses--277
Ormolu--199,200
Pails--66-68
Pedestal--291
Pen Staff Holder--239
Perfume Bottle--275
Pitcher--245
Planter (see Jardinieres)
Plaques--298,299
Port Hole Cover--160
Pressure Gauge Tester--154
Propeller--165
Pumps--139,140
Reading Stand--290
Rear View Mirror--112
Rub Board--87
Safety Deposit Box--230
Salesmen's Samples--10,11,59
Samovar--98
Scales--115-125
Sconces (see Wall Sconces)
Scoops--84,119
Sextant--170,171
Shells--155-157
Ship's Bell--162
Ship's Lamp--184
Ship's Reflector--164
Shoe Shine Stand--135
Shotgun Shell Crimper--136
Shower Heads--205,206
Signal Cannon--167
Slipper Boxes--53,54
Soap Dish--207,208
Soap Dish and Cup Holder--207

Sponge Holder--208
Spoons--80
Starboard Light--180
Steam Whistle--152
Stool--292
Surveyor's Instrument--131
Tea Caddies--85
Tea Pots--93-95
Telephones--294,295
Telescope--130,131
Test Tube Holder--138
Thermometer--151
Thimbles--276
Toasting Forks--45
Tobacco Jar--281
Toilet Tissue Hanger--210
Tool Rack--132
Towel Rack--209
Trays--96,104,106,296,297,300,301
Trench Art--156
Triplex Lens--179
Trivets--105,107-110
Trouble Light--113
Truck Tire Air Gauge--114
Umbrella Stands--304-306
Vase Holder--267
Vases--262,265,268,269
Voltage Meter--153
Wall Pocket Vases--260,261,264,266
Wall Sconces--16-19
Watering Can--92
Weights and Measures--115,116
Wine Cooler--100
Yacht Buoy Light--161
Yacht Tie-Downs--176,177
Yacht Wheel--163

Price Guide

Cover Photograph 86.00

PLATE 1 . . . . (pair) 250.00
PLATE 2 . . . . 195.00
PLATE 3 . . . . (pair) 300.00
PLATE 4 . . . . (pair) 2600.00
PLATE 5 . . . . (pair) 225.00
PLATE 6 . . . . (pair) 275.00
PLATE 7 . . . . (pair) 435.00
PLATE 8 . . . . (pair) 235.00
PLATE 9 . . . . 110.00
PLATE 10 . . . (pair) 18.00; two-light
                            holder, 6.00
PLATE 11 . left, 18.00; right, 20.00;
                            center, 38.00
PLATE 12 . . . . (pair) 47.50
PLATE 13 . . . . (pair) 550.00
PLATE 14 . . . . (pair) 110.00
PLATE 15 . . . . (each) 187.50
PLATE 16 . . . . (each) 300.00
PLATE 17 . . . . (each) 65.00
PLATE 18 . . . . 285.00
PLATE 19 . . . . 1200.00
PLATE 20 . . . . (pair) 550.00
PLATE 21 . . . . 650.00
PLATE 22 . . . . 145.00
PLATE 23 . . . . 55.00
PLATE 24 . . . . 150.00
PLATE 25 . . . . 75.00
PLATE 26 . . . . 125.00
PLATE 27 . . . . 300.00
PLATE 28 . . . . 195.00
PLATE 29 . . . . 495.00
PLATE 30 . . . . 225.00
PLATE 31 . . . . 195.00
PLATE 32 . . . . (pair) 150.00
PLATE 33 . . . . (pair) 225.00
PLATE 34 . . . . (pair) 300.00
PLATE 35 . . . . (pair) 650.00
PLATE 36 . . . . (pair) 600.00
PLATE 37 . . . . (pair) 175.00
PLATE 38 . . . . (pair) 375.00
PLATE 39 . . . . (pair) 235.00
PLATE 40 . . . . 650.00
PLATE 41 . . . . 55.00
PLATE 42 . . . . 80.00
PLATE 43 . . . . 135.00
PLATE 44 . . . . 35.00
PLATE 45 . . . . (each) 25.00
PLATE 46 . . . . (set) 175.00
PLATE 47 . . . . 165.00

PLATE 48 . . . . 360.00
PLATE 49 . . . . 175.00
PLATE 50 . . . . 85.00
PLATE 51 . . . . 135.00
PLATE 52 . . . . 175.00
PLATE 53 . . . . 75.00
PLATE 54 . . . . 70.00
PLATE 55 . . . . 315.00
PLATE 56 . . . . 630.00
PLATE 57 . . . . 525.00
PLATE 58 . . . . 150.00
PLATE 59 . . . . (each) 24.00
PLATE 60 . . . . 45.00
PLATE 61 . . . . 200.00
PLATE 62 . . . . 270.00
PLATE 63 . . . . 110.00
PLATE 64 . . . . 250.00
PLATE 65 . . . . 210.00
PLATE 66 . . . . 585.00
PLATE 67 . . . . 75.00
PLATE 68 . . . . 125.00
PLATE 69 . . . . 78.00
PLATE 70 . . . . 125.00
PLATE 71 . . . . 125.00
PLATE 72 . . . . 125.00
PLATE 73 . . . . 125.00
PLATE 74 . . . . 75.00
PLATE 75 . . . . 50.00
PLATE 76 . . . . 60.00
PLATE 77 top, 75.00; bottom, 55.00
PLATE 78 . . . . 65.00
PLATE 79 . . . . 300.00
PLATE 80 . . . . (each) 10.00
PLATE 81 . . . . 75.00
PLATE 82 . . . . 90.00
PLATE 83 . . . . 45.00
PLATE 84 . . left, 75.00; right, 55.00
PLATE 85 . . . . (each) 150.00
PLATE 86 . . . . 125.00
PLATE 87 . . . . 30.00
PLATE 88 . . . . 37.50
PLATE 89 . . . . 225.00
PLATE 90 . . . . 140.00
PLATE 91 . . . . 45.00
PLATE 92 . . . . 52.00
PLATE 93 . . . . 450.00
PLATE 94 . . . . 690.00
PLATE 95 . . . . 275.00
PLATE 96 . . . . (set) 200.00
PLATE 97 . . . . 45.00
PLATE 98 . . . . 450.00

PLATE 99 . . . . 900.00	PLATE 151 . . . . 29.00
PLATE 100 . . . . 42.00	PLATE 152 . . . . 250.00
PLATE 101 . . . . 15.00	PLATE 153 . . . . 12.00
PLATE 102 . . . . 45.00	PLATE 154 . . . . 105.00
PLATE 103 . . . . (pair) 125.00	PLATE 155 . . . . 12.00
PLATE 104 . . . . 45.00	PLATE 156 . . . . 250.00
PLATE 105 . . . . 85.00	PLATE 157 . . . . lamp, 67.50
PLATE 106 . . . . 47.50	shell alone, 25.00
PLATE 107 . . . . 125.00	PLATE 158 . . . . 185.00
PLATE 108 . . . . 120.00	PLATE 159 . . . . 19.50
PLATE 109 . . . . 180.00	PLATE 160 . . . . 600.00
PLATE 110 . . . . 75.00	PLATE 161 . . . . 30.00
PLATE 111 . . . . 110.00	PLATE 162 . . . . 850.00
PLATE 112 . . . . 225.00	PLATE 163 . . . . 350.00
PLATE 113 . . . . 24.00	PLATE 164 . . . . 350.00
PLATE 114 . . . . 35.00	PLATE 165 . . . . 60.00
PLATE 115 . . . . 100.00	PLATE 166 . . . . 225.00
PLATE 116 . . . . 165.00	PLATE 167 . . . . 850.00
PLATE 117 . . . . 275.00	PLATE 168 . . . . 385.00
PLATE 118 . . . . 30.00	PLATE 169 . . . . 300.00
PLATE 119 . . . . 75.00	PLATE 170 . . . . 750.00
PLATE 120 . . . . 200.00	PLATE 171 . . . . 750.00
PLATE 121 . . . . 137.00	PLATE 172 . . . . 1250.00
PLATE 122 . . . . 75.00	PLATE 173 . . . . 135.00
PLATE 123 . . . . 87.50	PLATE 174 . . . . 150.00
PLATE 124 . . . . 125.00	PLATE 175 . . . . 150.00
PLATE 125 . . . . 95.00	PLATE 176 . . . . 25.00
PLATE 126 . . . . 1250.00	PLATE 177 . . . . 65.00
PLATE 127 . . . . 1300.00	PLATE 178 . . . . 18.00
PLATE 128 . . . . 1025.00	PLATE 179 . . . . 350.00
PLATE 129 . . . . 75.00	PLATE 180 . . . . 325.00
PLATE 130 . . . . 1750.00	PLATE 181 . . . . 225.00
PLATE 131 . . . . 900.00	PLATE 182 . . . . 95.00
PLATE 132 . . . . 55.00	PLATE 183 . . . . 220.00
PLATE 133 . . . . 20.00	PLATE 184 . . . . 75.00
PLATE 134 . . . . 52.00	PLATE 185 . . . . 45.00
PLATE 135 . . . . 125.00	PLATE 186 . . . . 25.00
PLATE 136 . . . . 125.00	PLATE 187 . . . . (each) 12.00
PLATE 137 right, 75.00; top, 49.00;	PLATE 188 . . . . 55.00
foreground, 38.00	PLATE 189 . . . . 18.00
PLATE 138 . . . . 5.00	PLATE 190 . . . . (each) 32.00
PLATE 139 . . . . 75.00	PLATE 191 left, 26.00; right, 48.00
PLATE 140 . . . . 85.00	PLATE 192 . . . . (each) 15.00
PLATE 141 . . . . 19.00	PLATE 193 . . . . left, 8.00;
PLATE 142 . . . . 75.00	center 42.00; right, 12.00
PLATE 143 . . . . 140.00	PLATE 194 . . . . . . (each) 4.00-6.00
PLATE 144 . . . . 200.00	PLATE 195 . . . . . (each) 10.00-20.00
PLATE 145 . . . . 35.00	PLATE 196 . . . . (each) 6.00
PLATE 146 . . . . 100.00	PLATE 197 . . . . (pair) 11.00
PLATE 147 . . . . 25.00	PLATE 198 . . . . (each) 5.00
PLATE 148 left, 45.00;center, 35.00;	PLATE 199 left, 26.00; right, 18.00
right, 75.00	PLATE 200 top, 12.00; bottom, 18.00
PLATE 149 . . . . 5.00	PLATE 201 top, 36.00 pair; center,
PLATE 150 . . . . 35.00	40.00 pair; bottom, 54.00 pair

PLATE 202 left, 8.00 each; center, 15.00 each; right, 12.00 each
PLATE 203 . . . . (each) 12.00
PLATE 204 left, 22.00 each; right, 5.00 each
PLATE 205 . . . . 135.00
PLATE 206 . . . . .55.00
PLATE 207 . . . . .50.00
PLATE 208 . . . . .87.50
PLATE 209 . . . . .65.00
PLATE 210 . . . . .32.00
PLATE 211 . . . . .90.00
PLATE 212 . . . . 145.00
PLATE 213 . . . . .90.00
PLATE 214 . . . . .35.00
PLATE 215 . . . . .32.00
PLATE 216 . . . . 135.00
PLATE 217 . . . . 250.00
PLATE 218 . . . . 250.00
PLATE 219 . . . . 350.00
PLATE 220 . . . . 890.00
PLATE 221 . . . . 3500.00
PLATE 222 . . . . 785.00
PLATE 223 . . . . 225.00
PLATE 224 . . . . 250.00
PLATE 225 . . . . 275.00
PLATE 226 . . . . 350.00
PLATE 227 . . . . 125.00
PLATE 228 . . . . 270.00
PLATE 229 . . . . .45.00
PLATE 230 . . . . .27.00
PLATE 231 . . . . .25.00
PLATE 232 . . . . .60.00
PLATE 233 . . . . 100.00
PLATE 234 . . . . 450.00
PLATE 235 . . . . (pair) 95.00
PLATE 236 . . . . (pair) 40.00
PLATE 237 . . . . 215.00
PLATE 238 . . . . 300.00
PLATE 239 . . . . 175.00
PLATE 240 . . . . 480.00
PLATE 241 . . . . .40.00
PLATE 242 . . . . 290.00
PLATE 243 . . . . .65.00
PLATE 244 . . . . 160.00
PLATE 245 . . . . .30.00
PLATE 246 . . . . .30.00
PLATE 247 . . . . . (each) 15.00-20.00
PLATE 248 . . . . . (each) 15.00-20.00
PLATE 249 . . . . (each) 5.00
PLATE 250 . . . . 195.00
PLATE 251 . . . . .40.00
PLATE 252 . . . . 150.00
PLATE 253 . . . . 675.00

PLATE 254 . . . . 750.00
PLATE 255 . . . . 395.00
PLATE 256 . . . . 900.00
PLATE 257 . . . . .30.00
PLATE 258 . . . . .35.00
PLATE 259 . . . . 175.00
PLATE 260 . . . . 140.00
PLATE 261 . . . . 350.00
PLATE 262 . . . . .55.00
PLATE 263 . . . . .30.00
PLATE 264 . . . . .15.00
PLATE 265 . . . . (pair) 295.00
PLATE 266 . . . . (each) 55.00
PLATE 267 . . . . .25.00
PLATE 268 . . . . .80.00
PLATE 269 . . . . .58.00
PLATE 270 . . . . 165.00
PLATE 271 . . . . 325.00
PLATE 272 . . . . 120.00
PLATE 273 left, 60.00; center, 80.00; right, 60.00
PLATE 274 . . . . .15.00
PLATE 275 . . . . .27.50
PLATE 276 . . . . . . (each) 2.00-4.50
PLATE 277 . . . . .65.00
PLATE 278 . . . . .22.00
PLATE 279 . . . . .47.50
PLATE 280 . . . . .40.00
PLATE 281 . . . . .10.00
PLATE 282 . . . . .55.00
PLATE 283 . . . . .17.50
PLATE 284 . . . . (pair) 80.00
PLATE 285 . . . . .19.00
PLATE 286 . . . . .66.00
PLATE 287 . . . . .18.50
PLATE 288 . . . . .60.00
PLATE 289 . . . . .12.00
PLATE 290 . . . . 225.00
PLATE 291 . . . . 895.00
PLATE 292 . . . . 150.00
PLATE 293 . . . . 125.00
PLATE 294 . . . . 295.00
PLATE 295 . . . . 250.00
PLATE 296 . . . . .45.00
PLATE 297 . . . . .52.00
PLATE 298 . . . . (each) 6.00
PLATE 299 . . . . .50.00
PLATE 300 . . . . .30.00
PLATE 301 . . . . .68.00
PLATE 302 . . . . .55.00
PLATE 303 . . . . 195.00
PLATE 304 . . . . 225.00
PLATE 305 . . . . 300.00
PLATE 306 . . . . .80.00